CITIES OF FARMERS

Briana,

Thank you for being a great friend, collaborator, and a supporter of the Farmers Market.

Nurgül F.
December 22, 2016

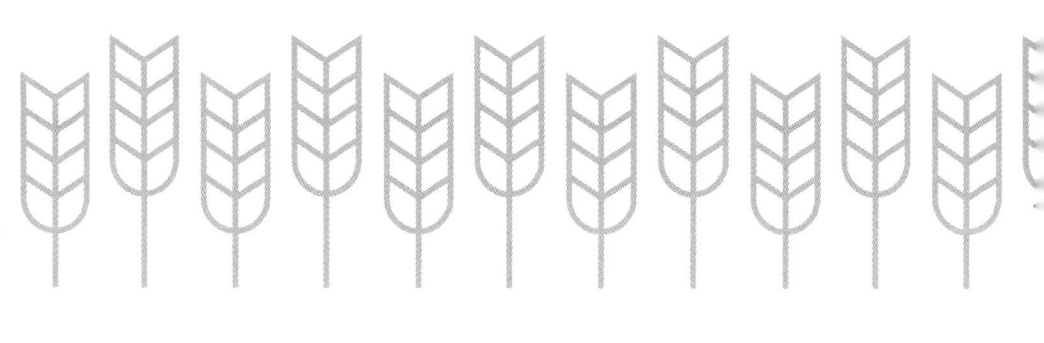

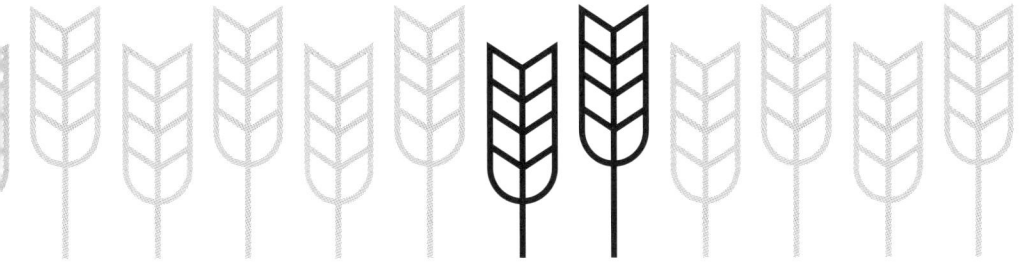

CITIES OF FARMERS

Urban Agricultural Practices and Processes

Edited by Julie C. Dawson and Alfonso Morales

University of Iowa Press, Iowa City

University of Iowa Press, Iowa City 52242
Copyright © 2016 by the University of Iowa Press
www.uiowapress.org
Printed in the United States of America

Design by Ashley Muehlbauer

No part of this book may be reproduced or used in any form or by any means without permission in writing from the publisher. All reasonable steps have been taken to contact copyright holders of material used in this book. The publisher would be pleased to make suitable arrangements with any whom it has not been possible to reach.

The University of Iowa Press is a member of Green Press Initiative and is committed to preserving natural resources.

Printed on acid-free paper

Library of Congress Cataloging-in-Publication Data
Names: Dawson, Julie C., editor. | Morales, Alfonso, 1961– editor.
Title: Cities of farmers : urban agricultural practices and processes / edited by Julie C. Dawson and Alfonso Morales.
Description: Iowa City : University of Iowa Press, [2016] | Includes bibliographical references and index.
Identifiers: LCCN 2016010366| ISBN 978-1-60938-437-1 (pbk) | ISBN 978-1-60938-438-8 (ebk)
Subjects: LCSH: Urban agriculture.
Classification: LCC S494.5.U72 C585 2016 | DDC 635.09173/2—dc23
LC record available at http://lccn.loc.gov/2016010366

To urban agriculturists, who inspire
us with their relentless pursuit of food justice

—JCD AND AM

CONTENTS

Acknowledgments xi

Foreword xiii
WILL ALLEN

Section 1. Introduction and Historical Antecedents 1

1. Cities of Farmers: Problems, Possibilities, and Processes of Producing Food in Cities 3
JULIE DAWSON AND ALFONSO MORALES

2. Food from Scratch for the Zenith of the Unsalted Seas: Creating a Local Food System in Early-Twentieth-Century Duluth, Minnesota 11
RANDEL D. HANSON

3. Municipal Housekeepers and the High Cost of Living: The Establishment of Gardening Programs and Farmers Markets by Grand Rapids Women's Clubs in the Early Twentieth Century 21
JAYSON OTTO

Section 2. Regulation 39

    4. Urban Ag' in the 'Burbs 41
       MEGAN HORST, CATHERINE BRINKLEY, AND KARA MARTIN

    5. Cultivating in Cascadia: Urban-Agriculture Policy
       and Practice in Portland, Seattle, and Vancouver 59
       NATHAN MCCLINTOCK AND MICHAEL SIMPSON

    6. Urban Agriculture: Composting 83
       LAUREN SUERTH

Section 3. Production 105

    7. Agroecology of Urban Farming 107
       ERIN SILVA AND ANNE PFEIFFER

    8. Lessons from "The Bucket Brigade": The Role of Urban
       Gardening in Native American Cultural Continuance 126
       MICHÈLE COMPANION

    9. Foregrounding Community Building in Community
       Food Security: A Case Study of the New Brunswick
       Community Farmers Market and Esperanza Garden 141
       LAURA LAWSON, LUKE DRAKE, AND NURGUL FITZGERALD

    10. Fumbling for Community in
        a Brooklyn Community Garden 159
       DORY THRASHER

Section 4. Distribution 177

    11. Food Hubs: Expanding Local Food to Urban Consumers . . . 179
       BECCA B. R. JABLONSKI AND TODD M. SCHMIT

12. Chicago Marketplaces:
   Advancing Access to Healthy Food 191
   ANNE ROUBAL AND ALFONSO MORALES

Section 5. Community Health and Policy Perspectives 213

13. The Coevolution of Urban-Agriculture
    Practice, Planning, and Policy 215
    NEVIN COHEN AND KATINKA WIJSMAN

14. Urban Agriculture and Health:
    What Is Known, What Is Possible? 230
    BENJAMIN W. CHRISINGER AND SHEILA GOLDEN

15. More Than the Sum of Their Parts:
    An Exploration of the Connective and
    Facilitative Functions of Food Policy Councils 245
    LINDSEY DAY FARNSWORTH

16. Embedding Food Systems into the Built Environment.... 265
    JANINE DE LA SALLE

    Conclusion 285
    References 293
    Index.. 329

ACKNOWLEDGMENTS

Alfonso would like to acknowledge his family: his grandparents and parents, who gave him a love of the land, and his spouse and son, Manuela and Cruz, who enjoy the fruits of his labor and who inspire his efforts. He would especially like to thank Lisa Jackson, Johanna Doren, and the students in his graduate classes in URPL 711, Food Systems and Marketplaces, particularly Jimmy Camacho, Danielle Smith, Alexander Brown, Erin Skalitzky, and Caisey Griffith.

Together Julie and Alfonso would like to thank Steve Ventura and the team of the Community and Regional Food Systems grant (USDA 2011-68004-30044, Ventura PI, Morales Co-I) and the Indicators for Impact grant (USDA 2014-68006-21857, Morales PI).

FOREWORD

WILL ALLEN

Urban food production has been practiced all over the world for a long time. In the Americas, it goes back hundreds of years. It is in our history. Aquaponics systems were being used in Peru centuries ago. There were the Victory Gardens during World War II, where urban Americans—with the active support and encouragement of the government—started growing food in their backyards and side gardens. During the great African American migration from the South to the North, many migrants planted gardens in their new backyards. They quietly grew their food, ashamed to talk about it because of their history of slavery on plantations and sharecropping during Reconstruction in the South. They were running away from something that was really painful.

All these methods of alternative agriculture were used because they were practical. African Americans in the North did not have a lot of disposable income even though their incomes were much higher than they had been in the South as sharecroppers. Victory Gardens were planted because of food rationing during World War II. The Incas in Peru used aquaponics to raise fish and grow high-quality produce in a harsh environment. Only recently have we started to return to some of these alternative agricultural methods.

I do not think we will be able to survive in the future without a dramatic change in the way we produce food. There are going to be 3.1 billion new people on earth in the next forty years, and we cannot even feed our population with good food now. The only way that we are going to be able to survive is to grow food closer to where people live—inside cities where there is a lot of vacant land. For example, the city of Milwaukee has 2,500 vacant lots. With all that vacant land, there is tremendous opportunity to grow food in the city.

CHALLENGES FACING URBAN AGRICULTURE

The challenge, however, is not a lack of land. It is that not enough farmers know how to grow food, or they do not have the right tools or compost to be able to plant on vacant land in cities. Land tenure is also a big issue. It does not make sense to spend thousands of dollars on land that you may not be able to stay on very long.

Another challenge for urban farming is that the regulation of urban agriculture by cities is changing all the time. Since I started doing this work, I have attended dozens of meetings with policy makers and city officials. This concrete work we have done in educating policy makers to show them that urban agriculture is a viable thing for cities to do has helped us change some of the policies inhibiting urban agriculture. The very existence of my organization, Growing Power, proves that this can be done, and this can be replicated in other cities all over the world. Urban agriculture is the future. We have to continue to inform policy makers about the viability of urban agriculture, and we also need to partner with like-minded organizations to increase our power in cities around the nation and to help create policies that are friendly to urban-agriculture efforts.

Yet another major problem that we face here in Milwaukee is water. Even though we are right on the shore of one of the Great Lakes, where there is a seemingly unlimited supply of water for agriculture, we struggle to provide water to our farm. Many of our growing sites are not connected to the city's water lines, so we have had to use fire hydrants. The city has been coming down on us for using the hydrants, but lack of water should not be something that stops you from farming in Milwaukee. We have overcome that for now by hauling water, but water policy is one of the issues we have to work on, not only here in Milwaukee but in other cities throughout the country.

Many urban-agriculture projects are facing similar problems in their respective cities. In Madison, Wisconsin, for example, where Growing Power has a presence in the form of the Resilience Center, there is a city ordinance that allows a hoop house to stand for only 180 days. It makes absolutely no sense. You have to take it down for the rest of the year, but we need production year-round. That is something that we are working on with Madison, but I am confident that we will win this battle because urban agriculture is here to stay. It is not a fad; we must have a sustainable food system in this country, and we are going to have to accomplish that with help from rural areas as well as urban areas. We just have to be patient in getting the policies to change at the city level.

Federal policy, on the other hand, seems to be embracing sustainable urban agriculture. The US Department of Agriculture (USDA) has a program that helps farmers pay for hoop houses, which is especially helpful for us in the northern United States. Trying to grow outside is very difficult. I do not care how good a farmer you are, you cannot beat Mother Nature. So with hoop houses we do the best we can. You are going to see more and more of these hoop houses, and eventually we will probably not even be growing outside anymore. That is how I see the future of urban agriculture. Hoop houses and greenhouses, aquaponic systems, sustainable energy, and getting off fossil fuels—that is all a part of the future.

I think that is what is exciting to me about this collection that Julie and Alfonso put together. It represents some of the history of urban food production, and it also represents some of what people are doing today. No one book can discuss everything, but this book is more than an introduction—it is a good representation of what people are doing around the country. I think readers will be interested in individual sections, and I think people will benefit from comparing the work across the different sections.

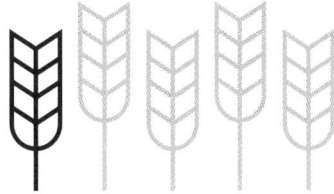

SECTION ONE
Introduction and Historical Antecedents

Urban agriculture, farmers markets, and regional food systems all feel new to our experience, yet they are not. Historians and other scholars remind us that practices we might consider "new" were once useful in advancing the goals cities, regions, and people had one hundred years ago and more. While we might quail at the language or at how people characterized each other, we can still see the threads common to our circumstances, and further, we can see how people of the past had goals we would recognize, even if not always completely similar to our own. Further, and important to this project, we learn from the following chapters how different types of food system activities were central to problem solving of the past, and to reaching to a future that we are now living. How will the future judge our works? Which of our activities will subsequent generations build upon?

CHAPTER 1

Cities of Farmers

*Problems, Possibilities, and Processes
of Producing Food in Cities*

JULIE DAWSON AND ALFONSO MORALES

Growing food in our cities affects every aspect of urban life. Urban agriculture can enhance local ecology, foster cohesive communities, and improve the quality of life for urban residents. However, these benefits can be hindered by tensions inherent in this emergent practice, requiring the reconstruction of personal habits and expectations for urban landscapes, as well as the reshaping of regulations to provide for such activities. Urban agriculture has the potential to improve the health and vitality of our communities, but realizing that potential often requires change to both the physical and political landscapes. In short, urban agriculture is one component of a larger urban food system, and to understand urban agriculture, it is necessary to understand the linkages between production, distribution, policies, and regulations. It is also important to bring out the relationship between peri-urban or rural agriculture and urban agriculture. While usually treated separately, these production systems are often interlinked in urban food systems, and opportunities exist to strengthen these linkages rather than hold rural and urban production in opposition.

In this book we advance the idea that urban agriculture is a unique production system, worthy of study on its own while at the same time sharing connections to other food system activities. We will also show how the practices of urban agriculture rest in a larger scaffolding of social and political organization. The authors of these chapters show that urban-agriculture systems are made up of interrelated activities that—as with all systems—involve interrelated parts, each with its own historical context and internal dynamics. The formula for success for any particular system varies with the context and the particular goals of those involved. The book also provides concepts and examples of activities related to how larger historical circumstances, social expectations and pressures, and other institutional spheres, like the law, influence people doing urban food system activities.

URBAN AGRICULTURE IN CONTEXT

The essays in this book explore the significant benefits and challenges of urban agriculture, mapping the complex economic, social, political, and ecological systems that make up a food biome in an urban environment. They show how growing food in vacant lots and on rooftops affects labor, capital investment, and human capital formation, and how urban agriculture intersects with land values and efforts to build affordable housing. Furthermore, they exemplify how municipal regulation of economic and urban-agriculture activities plays a key role in whether urban agriculture can flourish, while research has demonstrated that community gardens and fresh produce offer significant improvements in community and individual health.

Our purpose in providing this collection is to expose students, practitioners, scholars, and urban policy makers to common examples, ideas, and language for discussing, implementing, and evaluating urban-agriculture production practices in their local contexts. The book is a comprehensive examination of urban-agriculture systems, exploring the history, regulation, production, distribution, and health benefits of urban food production. This collection seeks to do more than describe aspects of our contemporary urban-agriculture practices. Our hope is that the content and discussion questions in each chapter will help make the lessons of the research accessible to a variety of different audiences and, perhaps most important, provide concepts and examples that render these lessons *actionable*.

The most meaningful and significant feature of urban or metropolitan agricultural production is its relative integration with other elements of the urban system. Urban food systems influence every human institution and practice. They influence the economy in terms of labor, capital investment, and productive activities, with implications for the value of surrounding housing and other land uses; policy in terms of ordinances and codes regulating different uses and activities, with implications for various uses of public space as in marketplaces; and society, in terms of the variety of health-related outcomes associated with producing, processing, distributing, and consuming food. Different people have different points of entry in their awareness of and interaction with urban-agriculture production. This has implications for both community organizations promoting urban agriculture and policy makers seeking to encourage particular outcomes in their food systems.

Likewise, urban-agriculture systems and food systems more generally are influenced by other social institutions, both historical and contemporary. Food system practices have been shaped by a combination of descriptive and normative concerns, with differing perspectives on both how to define urban agriculture and what purposes it should serve. Here our goal is to describe some of these connections, showing how producing food implicates culture, health, and law, while at the same time demonstrating that food production is determined by the *producers* in interaction with participants from other institutions. All these activities are practiced in context, with constraint and freedom being experienced in turn, and we urge readers to look beyond the specifics of the examples in order to understand the interconnections between various practices, to see different interpretations of similar words, and to recognize the profound influence of interaction and relationship, and of goals and hopes people have for their activities.

Many previous works have discussed urban food practices in the industrialized world, and there is a growing body of peer-reviewed journal articles (and new journals) on this subject, as well as a vast literature in the popular press. This volume focuses on the United States and provides historical examples of urban-agriculture production, as well as contemporary examples of urban agriculture in practice. Authors in this volume celebrate the successes but also advance examples that should give us pause to consider the intricacy of production agriculture in built environments of great social complexity. This

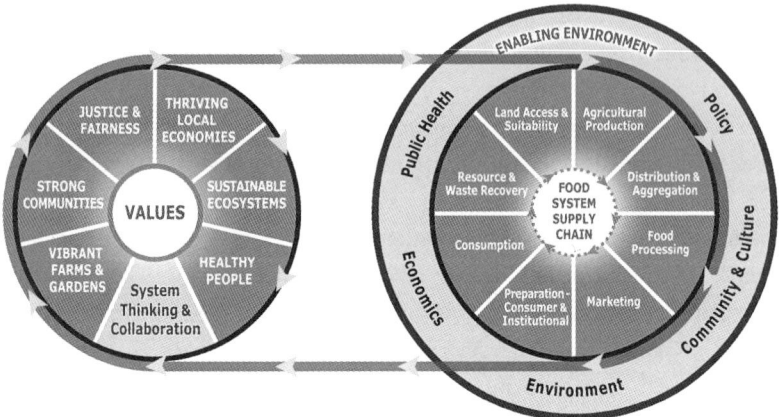

Figure 1. Community food system framework, developed by the Community and Regional Food Systems Project, http://www.community-food.org/.

volume makes a unique contribution to the literature by elaborating on key features of urban agriculture while at the same time providing a synthetic view across important subsystems.

Urban agriculture is only one aspect of the larger urban food system, and that food system shares connections with larger rural and global food systems. Thus, a food systems framework (fig. 1) is useful for visualizing the inherent complexity of urban-agriculture systems. It exposes the many tensions within and between systems and reveals the many possible purposes and benefits.

This version of the community and regional food systems framework has two sets of nested circles, the left side being normative and categorizing some of the values with which people imbue their food-related activities and attitudes and the right side being descriptive and categorizing the activities associated with the food system. The framework illustrates how food system activities (the inner circle on the right) are related to components of the *food system environment* (the outer circle on the right). The framework circles are nested "dials" whose rotation produces new relationships with different components of the food system environment. Each "wedge" in the dial on the left can be analyzed in the context of the whole food system environment or in relationship with one component of the food system environment. The chapters in this book can be considered as a series of successive "wedges" that describe distinct combinations of urban food system production as they relate to activities in the environment. Likewise, the circles on the left represent the normative dimension, or the values prevailing in society and its subgroups. The two sets of circles drive each other, as you will read in many of the chapters that follow and discuss how activities and values inform each other. Our focus is on issues in urban-agriculture production and how these relate to different aspects of the food system environment, as well as the values driving or reflected in the activities the chapter authors describe.

OUTLINE OF THE BOOK
Introduction and historical antecedents

This introductory section includes this overview as well as two chapters about the history of urban food systems. It should be no surprise that today we are rediscovering what was once useful. The lessons of the past speak clearly to us. Both chapters discuss urban food production in terms of community and cultural relations as well as health and nutrition. But perhaps more significantly, the authors recognize and reveal to us the economic, political, and social ideas that people associated with food production in the early twentieth century. The notion of a "prosperous" person, city, or society is related to connections with the soil in ways we find familiar but embrace today in different ways.

In chapter 2, Hanson shows urban food production and nutrition in the context of larger food system interests, policies, and quandaries, reminding us that historical urban food systems were not necessarily local by default, and that providing high-quality food to city residents has always required many organizations and subsequent collaboration between policy makers, planners, and citizens. The current discussion around relocalizing food production in urban areas is not new, but the context has changed. Our dominant food system now brings fresh produce to cities year-round, often at both an ecological and economic cost, and this system perpetuates many of the societal and economic inequalities seen in other domains.

In chapter 3, Otto points up the importance of the larger social context and how gardening was once seen as a tool of elites to improve the prospects of marginalized populations. His work shows how women and community organizations mobilized urban food production on behalf of disadvantaged community members. They did so in ways that made organizations more inclusive but did not always help transform gender roles or race relations. This chapter shows the degree of economic and racial paternalism seen in early urban-agriculture efforts, which existed alongside strong goodwill and desire for change on the part of organizers, and makes us aware that the same tendencies may be present in current efforts to promote urban agriculture as a solution to food insecurity. While urban-agriculture systems might be seen as an easily implementable solution to social problems, such an attitude betrays an ignorance of our organizational environment and the skeins of habits, practices, and relationships in which we all act. These historical chapters have lessons for us even as they foreshadow our contemporary concerns.

Regulation

One view of law is how it constrains our practices, but another is how law and regulation can enable activities that help achieve our common purposes. This section provides a vocabulary of law and policy and shows how people have used those ideas to advance urban-agriculture systems. The tensions that surrounded and continue to surround class and race initially led cities to discourage and regulate agriculture out of urban areas. Current students of urban food systems should prioritize learning this history so that we do not idealize the past or repeat past mistakes.

Regulations concerning urban food production are often invisible, spread over many divisions of government, or written in relation to other activities and then applied to food production. They often have not had dedicated attention to their development. Despite this disorganization, regulations and policies are central to urban food production and its related practices (e.g., composting), and they are some of the factors that most differentiate urban production from nonurban production. Chapters in this section reveal the connections between policy, waste recovery, and production.

Production

Urban food production practices take many forms. Land access is a critical component for urban agriculture, but the lack of secure land tenure has not prevented a significant amount of food production in cities. Through the examples in this section, we recognize that land, in many cities, has high value and is often available for only temporary use, making investments in soil improvement or infrastructure development difficult. The work in this section provides a rich vocabulary for thinking about urban food production.

Additionally, and importantly, these chapters examine production and organizational structures as well as the myriad relationships that form between production, management, and community, ultimately suggesting that some production practices are producing more than food—they are producing a network of civic agriculture. The lessons from this section exemplify the potential of urban-agriculture production and may be particularly useful to planners and advocates of urban agriculture seeking to improve distribution in areas of low food access.

Distribution

The chapters in this section relate production practices to consumers through showing how food system activities integrate production, distribution, and, in particular, logistics. At a time when many communities are considering food hubs and farmers markets, this section provides examples of the benefits that result from successful regional food distribution networks. The chapters in this section provide examples of distribution systems expanding the availability of local food and providing some of the efficiencies of larger-scale distribution without losing the connection between consumers and producers, thus demonstrating a successful alternative to the dominant large-scale system of food distribution.

Perhaps the most important contribution of this section is producing a research agenda that aligns research and measurement efforts with a meaningful understanding of food access specific to particular populations. Urban planners and policy makers often focus on brick-and-mortar stores for increasing food access, but this section examines other strategies that may be as effective and more sustainable over time.

Community health and policy perspectives

The final section provides an overview of the current understanding of urban agriculture and public health, and the interrelationships between policy, urban food production, the built environment, and population health. Various practitioners and audiences will find these chapters useful in how they use the language of various disciplines to show how to design health interventions, tailor evaluations, or form reasonable expectations about the effects of urban agriculture in their own communities. Included are case studies that review approaches and discuss best practices, in order to provide options for practitioners seeking to improve community health through the built environment, urban agriculture, and organizations such as food councils.

The interconnectedness of the issues involved in food systems is clear from the multiple topics that each chapter addresses. While this book is centered on urban food production, it is impossible to treat production in isolation. Throughout these chapters we see the opportunities in urban agriculture though innovations in production, policy, and community engagement. Food systems do not fit into neat categories, and they span multiple dimensions of urban planning and urban-rural connections. This is what makes them both exciting to develop and difficult to change quickly. This volume attempts to describe historical trends and document current innovations in urban agriculture with the goal of engaging multiple audiences in a discussion of the broader issues surrounding urban food systems and planning. It is with enthusiasm that we invite you to discover the many manifestations of urban agriculture and its relationship to healthy cities and citizens.

CHAPTER 2

Food from Scratch for the Zenith of the Unsalted Seas

Creating a Local Food System in Early-
Twentieth-Century Duluth, Minnesota

RANDEL D. HANSON

How do you create a locally harvested food system for a city of one hundred thousand? This question is being asked in many cities and regions across the United States. It was also an urgent local question a century ago.

Indeed, across the United States a century ago, public and private concerns were scrambling to get a handle on the haphazard process by which nature was transformed into edible human culture within rapidly urbanizing America. This was a chaotic, wasteful, and powerfully transformative period, with rural populations shifting into cities as the primary engine for economic activities shifted from agricultural to industrial (Tangires 2003; Cronon 1991; Danbom 1979).

The rapid growth of industrial cities forced an emerging municipal responsibility for the various inputs and outputs of this emergent urban life (Melosi 2008; Tarr 1984). Public and private city planners in the late nineteenth century began to reflect on and intervene in this laissez-faire urbanization, including how to procure ample food of adequate quality and reasonable cost to citizens (Morales 2000; Vitiello and Brinkley 2013). As was the case in many communities (see, for example, Jayson Otto's discussions of these issues as they related to Grand Rapids, Michigan, in the following chapter), it became apparent that leaving the issue of food to "the market" was wholly inadequate to meet the needs of the emergent society from any number of perspectives. Progressive-era politicians and citizens began to collaborate in planning for the needs of cities and their inhabitants, creating solutions as they were then defined. These histories of civic engagement with our food system by city governments, business organizations, and citizen groups represent a fascinating window into our past just as they help us think about our challenges and barriers for creating more desirable food systems within contemporary society.

While there were general issues that characterized the food challenges of early-twentieth-century industrial cities, many communities faced unique problems. The challenges faced by Duluth fell primarily into the latter category. Indeed, early-twentieth-century Duluth found itself in a food systems quandary. Situated on the western tip of Lake Superior amid vast, thick northern forests, the city was growing rapidly with the immense wealth garnered from exploiting the region's then-abundant natural capital. Timber from surrounding forests was being clear-cut and hacked into lumber to build the cities southward; the very rich and easily accessible iron ore of the Range was being gouged out and railroaded to Lake Superior docks in Duluth and elsewhere, filling ships and bank accounts; and grain from the newly plowed midwestern prairies and plains was being brought to port for shipping eastward, leveraging the ship canal and ever-improving harbor facilities for this zenith point in North America for oceangoing vessels. New steel plants were being built, and countless spin-off and allied manufacturing, supply, and production companies were proliferating in an urban-industrial frenzy. Nearly tripling in population across two decades, Duluth experienced a phenomenal rate of population growth that was greater than that of New York or Chicago in 1910, and local boosters fantasized that Duluth would become the North American hub as infrastructure developed (Van Brunt 1921). As a result of this combination of abundant raw material, labor, and natural amenities,

Duluth hosted more millionaires per capita at this point than any other city in the United States. These were heady times in Duluth, and the city fathers were indeed filling their plates.

Although the city was rapidly growing, more than eighty thousand Duluthians lived for the most part on the narrow 24-mile strip of land hugging the western Lake Superior shore. The surrounding region was very sparsely populated save for the booming and busting mining and timber towns spread across the hinterlands. Eugene Van Cleef, a geographer at Duluth State Normal School (which would become the University of Minnesota–Duluth), worried in an article published in the *Bulletin of the American Geographical Society* in 1912 that the "permanence" of Duluth was threatened by the lack of an agrarian base, warning that "mineral resources alone do not invite a large population; they must be accompanied by food to support the people who market them" (Van Cleef 1912). More to the point, business leaders of Duluth were worried about attracting the important middlemen and women to run the businesses that were proliferating: poor-quality food, it was feared, would hinder their importation. And they were worried as well about the prices of food, which in cities around the United States were reaching all-time peaks, often taking between 40 and 60 percent of an average family's income (Donofrio 2007). Riots were sparked by this situation in New York and elsewhere, and the strong Duluth labor community (and its diverse political ideas and aspirations that challenged the status quo) was seen by industry captains as potentially fomenting local protest around food prices (Hudelson and Ross 2006). Given the lack of any local food supply, the availability and price of food in Duluth were indeed problematic.

The Duluth Commercial Club (a forerunner of a chamber of commerce type of organization) was at this point a powerful civic and political organ that assembled and channeled the business interests of Duluth, and its members began to consider the necessity of proactively building a local food supply (Stockbridge 1913). At the turn of the century, some of the wealthy members of the club had purchased clear-cut land for their summer homes beyond Skyline Drive, which ran along the top of the 1,800-foot drop into the Lake Superior basin. On these lands they began to dabble in agriculture and animal husbandry. As these "city fathers" carried out their projects, they realized both the potential of agriculture in the region and the difficulties, including dealing with the stumps of clear-cut trees, the new secondary growth that quickly sprouted up, and the rising prices of arable land in the area.

Taking all these issues into account, Duluth Commercial Club members sketched out a plan to jump-start a food system from scratch, including production and distribution components, to supply fresh produce to area restaurants, grocers, and households ("March of the Cities" 1911). Club secretary Major Eva created an agricultural subcommittee of its Public Affairs Committee, and one of their first actions was to hire Mr. A. B. Hostetter, a lifelong farmer and long-term teacher of farmers in the agricultural institutes of Illinois. Hostetter was turned loose with his considerable experience and sufficient club resources to pull together the educational, public relations, and networking elements to spark a local food system. Other club directors, including Charles Craig, owner of the Jean Duluth Farm and all-around entrepreneur, went to work on creating structures that could channel the developing public and private interests around food and agriculture.

After sizing up the situation, Hostetter approached the Duluth public schools to embrace agricultural education, but they demurred. Undaunted, Hostetter approached the YMCA, which began offering classes in poultry production in 1910; by 1911 the "Y" had added gardening classes, integrating a teacher for each of the twenty public schools in the city (Stockbridge 1913). Hostetter also worked with the Duluth Homecroft Association (DHA), a local arm of the national Homecroft movement, designed to encourage local self-sufficiency and healthy living (Garvey 1978). As a "model city" in this movement, Duluth boasted the founding in 1909 of Homecroft Park, which sold one-acre lots to area residents for a back-to-the-land urban lifestyle. Hostetter harnessed the energy of this movement by partnering with the DHA, which began to offer courses in cooking local produce, preserving foods, and managing the vagaries of such enterprises. Various churches, fairs, and community gatherings were encouraged to hold friendly competitions over the fruits and vegetables of these labors to generate greater interest. And the prized specimens were also brought to state fairs in Saint Paul, New York City, and other places to boost the image of agriculture in the region and attract potential farmers.

Mr. Hostetter and other Duluth Commercial Club members also leveraged their networks and the growing food needs of US Steel and its employees by partnering with the various railroad companies in the region, each of which had excess lands adjacent to its tracks (Stockbridge 1913). Together they crafted plans to create farms along the tracks, bunched into groups that would become small towns connected to the nearest train stop, which could serve as a portal for produce bound for urban destinations. To help grow these small centers for agricultural production, Hostetter created "educational trains" in which agronomic experts in seeds, produce varieties, production methods, management expertise, and so forth would travel on appointed days, stopping at each town to dispense their knowledge, praise, encouragement, and institutional support. Free seeds were distributed to town children, who were encouraged to compete with each other to grow the best produce, with the winners garnering prizes that the club also dispensed.

But problems in boosting a food system also existed because of a lack of access to lands closer to the city that could be agriculturally productive and affordable. Indeed, given the rapid population growth and the craggy landscapes along Lake Superior, land was quite expensive and arable land was scarce. How could you justify farming on land close to the city that was so expensive? To address this problem, several Duluth Commercial Club members, led by Charles Craig and mining lawyer and future University of Minnesota regent John G. Williams, founded the Greysolon Farms Company in 1910 (Mattocks 1911). The Greysolon Farms Company land occupied a mile-square area on Duluth's northern urban edge near Jean Duluth and Martin Roads; the land was developed as small farms ranging from one to fifteen acres for both rental and sale to workers, truck farmers, and distant farmers who might be coaxed from elsewhere to relocate. Craig and colleagues devised long-term financial agreements amenable to people to both rent and purchase land from which stumps were removed, and they created another, less expensive track for those who were willing to remove such obstacles to farming themselves. And as part of the deal, the Greysolon Farms Company would help people learn the skills of "intensive cultivation, market gardening, and dairy farming under the most modern scientific conditions" so they could make profits sufficient to justify purchasing the lands (and fulfill the food-supply ends of the club). The Greysolon Farms Company quickly took off, renting and selling agricultural lands for home and market production. Educational courses were held on the Greysolon Farms Company lands, organized by Hostetter, helping the aspiring farmers gain the necessary skill sets to produce for nearby markets. The creation of the Greysolon Farms Company was also not coincidentally commercially successful, creating profits for investors by adding value to cutover lands by removing stumps and getting the lands into cultivatable condition.

The University of Minnesota was also interested in inserting itself into the formal development of an agricultural infrastructure in the western Lake Superior region as part of its broader land grant mission (Thompson 1938, 1954). In 1911, the Minnesota state legislature authorized the Board of Regents to come to Duluth to seek lands that could support an experimental station akin to others that it was creating around the state. The Greysolon Farms Company lands were widely seen as the best farmland in the immediate Duluth area. The university negotiated hard with Greysolon's owners, drawing out the arguments for over a year, but eventually the university purchased some 240 acres at Greysolon's asking price and founded the Northeast Demonstration Farm and Experimental Station. Bolstering Greysolon Farm's activities, this new station quickly ramped up its operations. By the spring of 1913, Superintendent Mark J. Thompson was hired and the farm quickly developed as a combination dairy, poultry, and truck farm. Although the "Great Fire of 1918" burned this area, it was a temporary setback: the Northeast Experimental Station (which soon became its official name) became an important piece of the agricultural architecture of the region as a site for demonstration, production, and education. Thompson remained a main force on the farm for several decades, contributing to the "golden years" of research and extension services in the region.

Seeding education and production lands were two key aspects of building a food system from scratch that were now set in motion, but distribution was also a problem. To address this problem, the Duluth Commercial Club worked with area farmers to found a Cooperative Produce Warehouse in west Duluth in 1910 to supply goods to city retailers ("Model Co-operative Marketing Association" 1911). This experiment soon ran up against stubborn economic realities: there were not enough farmers bringing produce to the warehouse to make it economically self-sufficient, and the Commercial Club, which was underwriting the project, soon grew dismayed with the ongoing financial losses and finally shut the doors. In the wake of the closure, the club worked with area farmers to create the Producer's Cooperative Market Association as a more diffuse organizational means to represent and boost the interests of area farmers in distribution issues. In addition, the city of Duluth founded the Duluth Farmers Market in 1912 to service private households (Stockbridge 1913; see also Morales 2000 on Chicago's Maxwell Street Market). This first iteration of the Duluth Farmers Market, regulated by the City Council, opened up shop in the armory, adding two additional satellite markets in other parts of the city. That first year twenty-five farmers used the market to sell produce, which was all locally harvested, and the Duluth Farmers Market has in one form or another remained a part of the city ever since.

In sum, an amazing amount of energy and organization was brought to bear on the creation of a local food system for Duluth in the early part of the twentieth century. For an interim period that lasted several decades, this bid to create a local food system worked: locally harvested produce began to flow into area outlets, people turned to farming as an occupation, and other distant farmers relocated here. This local food system grew throughout the 1920s and 1930s, and vegetables like potatoes, brassicas, celery, and lettuce became staples that were grown in large fields sufficient to supply locally and ship elsewhere. Small-fruit production, particularly raspberries, was also robust enough to not only supply the region but also ship refrigerated train-car loads to Milwaukee, Minneapolis, Chicago, and Omaha. Simply put, regional food production thrived.

But as it did with so many aspects of US society, the advent of World War II signaled a profound change for the Duluth local food system. For one thing, the war effort demanded that as many people as possible work in activities related to iron ore and steel. In addition, the wage-oriented consumer society that flourished after the war continued the movement away from agriculture in the region. Small farms developed over the previous several decades were abandoned, and today we see these overgrown places all around the area. By the 1950s, larger-scale commercial farming across the United States began to edge out small-scale producers en masse, and regional and international specialization and development created the basis for the global industrial food system (Thompson 1959). Corporate farming became an increasing norm, as agriculture become vertically integrated into global food corporations. Farming in northern Minnesota ebbed steadily given the ever-cheapening cost of industrial food produced by externalizing so many of the trust costs of both production and distribution methods. Suburban sprawl began to creep into the richer agricultural lands north of the city. By the mid-1970s, a regional food infrastructure seemed too outmoded if not already gone, and the Northeast Experimental Station was closed in 1976, signaling a tardy ceremony for the ending of a local food system in the western Lake Superior region. And if this dirge was not heard, the small-farm crisis of the 1980s drove nails into the proverbial coffin of smaller-scale farming in the region, the state, and across the country (Hurt 2002; Lyson 2004).

In the wake of this industrialization of farming, nascent organizations designed to support small-scale sustainable farming and gardening began to appear across the United States, inspired by the resilient voices of people like Rachel Carson, Wes Jackson, Wendell Berry, Barry Commoner, and others. Community-based gardening in Duluth began to take shape in the late 1970s, and in 1981 the Duluth Community Garden Program was formally founded. Food cooperatives appeared in the 1970s, including Duluth's Whole Foods Co-op, which continues to expand into the present. The Land Stewardship project was created in Minnesota in 1982, and in 1988 the Sustainable Farming Association was created. These local and regional organizations have sought to continue a home garden and small-scale agriculture practice amid a fast-food and industrial agriculture that has systematically de-educated people about food and the skills of farming and gardening. Now that the health, community, and ecological bills of the externalizing system of industrial agriculture are coming due in ways that can no longer be ignored, these community and regional organizations resonate with greater authority and importance as we seek to bring their visions from the margins to the center (Reganold et al. 2011; Syring 2012; Stark, Abazs, and Syring 2011).

To paraphrase the English scholar and activist Stuart Hall, "hegemony is hard work." The hegemony of the global industrial food system is both powerful and rigid: we partake in its reproduction with an unnerving knowledge of its destructive wake. How can we use this knowledge to build a healthier food system for individuals, communities, and the landscape? Looking back on this largely successful early-twentieth-century transformation of the Duluth food system, we see some intriguing pathways. The people who organized the changes thought big and systemically, they integrated people and organizations across sectors, and they leveraged powers beyond Duluth that had interests in the city. How can we use their story as we wrestle with smart decline from an industrial paradigm with eyes wide open in optimism for the possibilities of a more sustainable future for ourselves and those who will find themselves on these same soils a hundred years hence?

In short, how can we work for the "permanence" of Duluth by laying the foundations for a sustainable food system? Time to put our shoulders to the wheel, as the arc of history does not get bent in a just and sustainable direction on its own.

DISCUSSION QUESTIONS

1. How was urban agriculture in Duluth one hundred years ago similar to urban agriculture today? How was it different?
2. "Rust Belt" cities like Duluth have been among the hardest-hit cities in the current economic recession. Given that urban agriculture originally developed in Duluth because of the economic hardships of many, could urban farms make a return to Duluth under the present circumstances? If yes, how would urban agriculture come into fruition?
3. Businesses, universities, and organizations in a similar vein to chambers of commerce were at the forefront of urban agriculture in Duluth. To what extent did urban agriculture develop either from the "top down" or the "bottom up"? How does this compare with urban agriculture today?
4. Consider the actions of Mr. Hostetter and the Greysolon Farms Company. What success did they have in northern Minnesota that could be replicated today? Did they make any missteps that those hoping to start urban-agriculture projects in their cities could hold as a cautionary tale?
5. How did the University of Minnesota bolster urban agriculture in Duluth at the turn of the last century? How do universities support urban agriculture today, and how has that changed over time?
6. In the 1950s, "Big Ag" began to overtake local food production, and by the 1970s Duluth's original food system was virtually dismantled. However, the Duluth Community Garden Program was founded in

1981. Given the rich history of urban agriculture, is urban agriculture still on the decline, or is it experiencing a renaissance?
7. Duluth was an exceptional urban-agriculture case because it involved collaboration from so many parts of society—particularly from business leaders. Can such collaboration occur today? Should it?

CHAPTER 3

Municipal Housekeepers and the High Cost of Living

The Establishment of Gardening Programs and Farmers Markets by Grand Rapids Women's Clubs in the Early Twentieth Century

JAYSON OTTO

On March 21, 1917, Eva McCall Hamilton was quoted on the front page of the *Grand Rapids Herald*: "The retail market will become an institution in Grand Rapids. The turning over of vacant city lots to householders is also a splendid idea. These things show the spirit of Grand Rapids." Hamilton, who would go on to be the first female senator in Michigan, was reveling in her victory won the previous evening at Grand Rapids City Hall. That night she had used her influence as a member of the mayor-appointed High Cost of Living Commission to overcome a filibuster blocking a $2,000 appropriation for the development of the first public retail market in Grand Rapids, Michigan.[1] At the same meeting, she helped push through a resolution that gave the city comptroller power to distribute over three hundred vacant lots in the city to individuals, families, or other groups of community members wanting to grow their own food (*Grand Rapids Press* [GRP], 20 March 1917, 5).

Public retail markets, which are what we would now consider "farmers markets," became an institution in Grand Rapids through the work of wealthy, upper- and middle-class women concerned with the living conditions of the working classes. By 1922, civic work had led to three city-operated markets located in working-class, immigrant neighborhoods. These markets were public services that alleviated the high cost of living caused by massive inflation, stagnant wages, and ultimately the war in Europe.[2] They were not warmly received by local grocers, however. Retail and wholesale grocers fought the idea of retail farmers markets for at least twenty years in Grand Rapids, protecting their near monopoly of fresh produce, which was supported through municipal ordinances.

Prominent Grand Rapids women like Hamilton were able to overcome the domination of grocers when they were given official space in local politics as consultants for issues of social welfare. Their interest in retail markets and public gardening programs stemmed from a general concern for "municipal housekeeping," an idea heavily promoted across the country by progressive female reformers such as Reverend Caroline Bartlett Crane of Kalamazoo, Michigan (Davis 2006; Rynbrandt 1997). She and other educated female leaders urged all women to actively work with local governments to promote health, cleanliness, and all things domestic. In Grand Rapids, one of their first and most successful campaigns was a school and home gardening program among children.

Gardening programs in Grand Rapids were widely supported through public and private collaborations, forging strong relationships between local business owners and politicians and groups such as the Ladies Literary Club. These associations led to female appointments on official working groups in the local government that ultimately influenced the establishment of the first retail farmers markets in Grand Rapids. Their endeavors are an early example of work around food production that reinforced the female social role while also politically empowering women. Such food-based social movements still tend to be gendered (Delind and Ferguson 1999), and, as Nevin Cohen and Katinka Wijsman have shown in chapter 13 with New York community gardens, female political empowerment is still an important ancillary to the explicit benefits of urban agriculture.

Unlike public gardening programs that enjoyed universal support, the public retail markets of Grand Rapids were contested public services whereby politically active women placed responsibility for feeding the city solely on the local government. City officials were reluctant to appropriate funds for a public service that was considered a detriment to local business owners. Nevertheless, the special interests of grocers were eventually eclipsed when club women were appointed by the mayor to the city's High Cost of Living Commission. Drawing from the momentum created through their garden programs and backed by reports from other cities, these women argued that Grand Rapids should operate retail markets as a way to contribute to wartime food production and fight the high cost of living. After an experimental period of a few years, the city purchased and developed permanent sites for the markets and hired a city employee to manage them.

This chapter explores the relationship between early public gardening programs and farmers markets in Grand Rapids during the first quarter of the twentieth century and also contributes to the literature on the work done by early-twentieth-century women's clubs. It begins with a description of fresh produce retailing at the turn of the twentieth century in Grand Rapids and describes how sales by farmers were eventually consolidated at a city-owned wholesale market that was inconvenient for residents wanting to purchase directly from farmers. The second section follows the work of women's clubs in their pursuit of improving the city through gardening. Their work resulted in city-wide school gardening clubs that were supported financially through partnerships with local, regional, state, and federal public institutions and local business organizations. The third section explains how the same women's organizations dealt with local grocers and gained the support of housewives to leverage appropriate funding for an experimental farmers market. The chapter ends with a personal experience with the last remaining original Grand Rapids farmers market and uncovers a disconnect between the market "experts" and the local reality, namely, their vision of the farmers market beyond its historical role of simply providing a public service for the people of Grand Rapids.

THE END OF DIRECT MARKETING IN GRAND RAPIDS

Similar to Duluth, Minnesota, described by Randel Hanson in the preceding chapter, Grand Rapids grew rapidly toward the end of the nineteenth century. It expanded from a town of about 3,000 people in 1850 to a city of over 110,000 by 1910, making it the forty-fourth largest city in the United States (US Census Bureau 1913, 899). This growth was stimulated through manufacturing, most notably by innovations in the office furniture market. The furniture industry and other manufacturing companies employed about twenty thousand wage laborers who were mostly Dutch and Polish immigrants living in congested ethnic enclaves (Kleiman 2006). As was the case in many other rapidly growing cites, purchasing food could be laborious and was usually placed on the shoulders of women (Deutsch 2010).

There were three main options for acquiring fresh produce in Grand Rapids: corner grocers, hucksters, and the early-morning downtown market. Of course, one could grow one's own vegetables. Indeed, Grand Rapids was unique in that home ownership was high. Many families had spaces and were encouraged to garden free from many of the prohibitory regulations on gardening found in some cities during the same period (Fitzgerald and Ely 1921; Horst, Brinkley, and Martin in this volume). However, families worked long hours to be able to afford home mortgages, and most did not have a refrigerator to store perishables or the resources and time to can. Therefore, households relied on the grocers and hucksters for their daily food supply, or they woke up early to catch a streetcar downtown to buy from surrounding farmers.

Unlike early-twentieth-century Duluth, Grand Rapids was well served by its agricultural surroundings, and a fairly efficient distribution system was already well established by the turn of the twentieth century. Scores of farmers descended on the downtown streets as early as midnight and congregated at Porter Block, directly in the center of Grand Rapids (GRP 23 September 1893, 1; 21 May 1921, 10). Most sales were made to the many jobbing and commission buyers whose grocery wholesaling business eclipsed that of Detroit (Grand Rapids Board of Trade 1888). These buyers were interested mainly in peaches and other high-dollar fruits that could be sold to distant markets. Vegetable growers, who were mainly small-scale Dutch immigrant farmers, were eventually segregated from the wealthier fruit growers to make it easier for the commission buyers and wholesalers to purchase fruits, and they were forced into less-desirable spaces through police coercion (GRP 31 August 1895, 4). Grocers and hucksters worked their way through these fruit and vegetable sections early and found the best deals, while householders shopped during the last couple of hours of the market.

By the mid-1890s, this market had spread over most of downtown, and the city government began looking for better locations at the urging of downtown business owners and fruit growers. A site was agreed on in 1896, and the market was moved to one of the islands in the Grand River in the spring of the next year. Although the purchase was decided through a popular vote, turnout was low in working-class neighborhoods and an appeal was made by aldermen, arguing that voters were not given a clear understanding that the market would be on the island (*Grand Rapids Common Council Proceedings* [GRCCP] 12 January 1891, 4 September 1894, 25 February 1895; *City of Grand Rapids Annual Report* 1897; GRP 6 October 1896, 4). Streetcar lines did not run to the new market and retailing was actively discouraged. Small-scale fruit farmers had petitioned for a retail market but were summarily ignored. The new market benefited large fruit growers as it was near the commission houses, and prime market stalls could be purchased for higher rents, as opposed to the first-come, first-served stall designations of the previous marketing arrangement (*Michigan Tradesman* [MT] 11 August 1896, 5; GRCCP 22 May 1893; GRP 18 April 1899, 3).

In addition to the new wholesale market, laws were put in place whereby farmers would be arrested if they sold on downtown streets. These laws were the result of pressure by downtown business owners, grocers, and wholesale merchants. Farmers were now forced to either sell at the wholesale market or bring their products directly to homes or retailers (GRCCP 22 May 1893; GRP 18 April 1899). They could not stop their wagon in the street to sell. This arrangement gave the city's three hundred retail grocers and sixty-plus hucksters a near monopoly over fresh produce sales in Grand Rapids neighborhoods (Polk 1893). Grocers had an advantage over the hucksters. They were able to influence how many of these licenses could be given to the mostly Dutch and Jewish food peddlers (*Grand Rapids Herald* [GRH] 30 July 1899, 8).

When food prices rose dramatically during the second decade of the twentieth century, national public discourse began questioning the honesty of middlemen and food retailers, and accusations of high markups and speculation were rampant. The "high cost of living" became the buzzword in Grand Rapids newspapers beginning around 1910, and many citizens started to question local purveyors of meat, milk, and fresh produce. The first prominent, vocal critic of the produce marketing system in Grand Rapids was Charles H. Leonard.

Leonard saw the establishment of the wholesale market as a direct attempt to monopolize retail produce sales. He stated in a letter to the City Council in 1911, "When the ground was originally purchased it was supposed that when it was improved the people would have access to it, and be able to trade directly with the farmers, but through some influence or other no steps have been taken to carry out the original plan. On the contrary the business has fallen into the hands of a lot of Jobbers who buy up the farmers' produce and then sell to the hucksters who divide up the city to rob the people" (Leonard 1911).

Leonard owned the massive Leonard Crockery Company, which was the sole manufacturer of cleanable, enamel-lined iceboxes, and he promoted retail markets as a way his workers could find reasonably priced food in Grand Rapids. He presented a plan for an official time set aside for retail sales at the wholesale market. "Pennies are so all fired scarce among the laboring men that I am convinced they will be glad to make the market trip even if they save but a single dime," he stated (GRP 17 May 1912, 21). His plan was popular among labor unions and in local editorials, but bitterly opposed by grocers and ridiculed in trade publications (Peters 1911; Zuller 1911; MT 29 May 1912, 8).

The city granted him the chance to set up an experimental retail market between 6:30 and 9:00 a.m. at the wholesale market site, complete with streetcar access from the neighborhoods. Unlike the successful start of the Duluth Farmers Market that same year, Leonard's experiment was a miserable failure. The grocers easily dismantled the retail market experiment by boycotting farmers choosing to sell to housewives (GRP 9 May 1912, 1). These boycotts continued for at least another five years, making it difficult for small growers to make a living selling produce in Grand Rapids. A small farmer commented on the state of the Grand Rapids wholesale market in 1916, "One of the greatest reasons the present market is a failure to the consumer is because if the farmer or gardener sells in small lots to the consumer who comes on the market to buy household supplies the grocers and hucksters will not patronize him. Often one would sell in small amounts but must refuse or a boycott would soon follow. This is the unwritten law of the present market system. Let the large grower or farmer sell to commission men, hucksters, and grocers, but give the small man a chance at a retail market" (GRP 8 December 1916, 6).

The first public retail farmers market was opened a year after this farmer's observation. However, it was not established explicitly as a benefit to small farmers. It was part of a larger movement that improved living conditions for working-class families by bringing the production and distribution of food closer to Grand Rapids residents, and it was spearheaded by women's clubs that had positioned themselves as the city's housekeepers.

MUNICIPAL HOUSEKEEPING AND PUBLIC GARDENING PROGRAMS

Heavily influenced by the City Beautiful campaign that was sweeping the country, Grand Rapids club women saw themselves as the housekeepers of the city, which was, as one writer of the time claimed, "of natural interest to women" (Haskins 1912, 4). The Grand Rapids Ladies Literary Club led the way, appointing a Civic Health and Beauty Committee in 1903, which supported various initiatives such as a small park system, beautification of streets, protection of current parks, and school-based gardens (*Grand Rapids Ladies Literary Club Minutes* [GRLLCM] 1903–1904). Beautifying the city was serious business, and these first-wave feminists were experimenting with novel ideas, most notably in the public and parochial schools. For example, a real issue in Grand Rapids was the amount of fresh air children breathed. Inside air was considered unhealthy, as evidenced by the experimental "open-air" classrooms in Grand Rapids, which kept the windows open even in the dead of winter (Michigan Association for the Prevention and Relief of Tuberculosis 1909). Gardens, too, were a way that students could get out into the fresh air.

The school gardens ended up being the foundation of the Civic Health and Beauty Committee's work. School gardens were seen as an extension of the classroom where students not only learned how to grow flowers and vegetables but also gained an appreciation for rural life and nature. Unlike the gardening programs in Duluth, which gave priority to food production and agricultural know-how, producing food was secondary to manual training and cooperative self-improvement. This was the case in many other cities establishing public gardening programs during the early part of the twentieth century (Lawson 2005). An added bonus was that the gardens would keep students out of mischief during summer vacation. In fact, gardens as a panacea for the problems of urban youth dominated the early discussions around school gardens in Grand Rapids. For example, one reporter wrote,

"One of the penalties of urban life is that the younger generation too often becomes removed from that close contact with the soil which makes for sane ideals and high-minded citizenship" (GRP 10 September 1910, 4).

The first school garden was an experimental plot started at the Second Avenue School in 1903. The school was located next to the rail lines, not far from the wholesale market. Second Avenue was a working-class school, most likely serving immigrant children. The superintendent of schools, William H. Elson, worked with the teachers and neighbors to have the school grounds plowed and planted with radishes, onions, lettuce, parsley, and potatoes. Gardening was seen as a way Second Avenue students could receive hands-on education while also supplementing their household food (*School Journal* 1 April 1905, 347–50).

A home gardening experiment was started by the Civic Health and Beauty Committee at the same time as the Second Avenue school garden and was inaugurated with a distribution of over three thousand seeds, mostly flowers, to elementary students (GRLLCM 1903–1904). Addresses of the students were recorded and inspections made throughout the summer. Prizes were offered for the best displays of flowers, and the Grand River Valley Horticultural Society offered an annual space for the best school garden displays at the West Michigan State Fair. By 1910, twenty schools were participating in seed distributions and school gardens. Local papers would report on the annual entourage of wealthy citizens such as physician Frances A. Rutherford and Eva Hamilton as they toured the city in automobiles, inspecting garden displays at schools and the home gardens of pupils (GRP 10 June 1905, 7; GRH 15 September 1905, 1; GRP 27 September 1905, 2). Letters written to the superintendent by student participants were published in the papers, and reporters were eager to interview participants to show the progress that had been made. Quotes like the following were common:

> At South Division street school one little negro boy proudly displayed a cornstalk ten feet high. He had no good soil in his own yard in which to plant the seeds so he borrowed some from the next-door neighbor. The corn grown under these difficulties was one of the finest stalks seen at any of the schools. . . . Many of the child gardeners worked under discouraging difficulties and were consequently proud of their suc-

cess. One boy when asked why he had no flowers or vegetables looked genuinely disappointed and, lapsing from school room English to street jargon, said: "Aw 'taint no use, every time I tried to raise a garden the things just begin to grow and we has to move. I'm awful sorry, lady, but 'taint no use." (GRP 9 September 1910, 2)

It is significant that African Americans were included in the report of school gardens, given that they made up less than 1 percent of the Grand Rapids population during this period. Both recent immigrants and African Americans were the topic of much lively debate in Grand Rapids, and their designation as "problems" reflected the racist attitudes of the time. The school gardening experiment initially focused on working-class, minority neighborhoods and the initiative may have been aimed at fixing these "problems." Another reporter commented, "One principal of a building in which are many children of foreign parentage says that while it was difficult to keep the school lawn unmolested the children now take personal pride in its appearance and have even contributed pennies to buy flowers, beside planting and cultivating the seeds and bulbs secured through the committee on civic health and beauty" (GRP 3 August 1907, 7).

The paternalistic nature of the work done by these women was explicit and did not go unnoticed. As one woman expressed, after going to a workshop a few years later where Eva Hamilton taught women how to conserve food, "It is almost laughable to hear one who knows nothing about poverty tell the poor how they should save" (GRP 17 March 1915, 6). Nevertheless, these women were able to leverage support from a wide variety of sources. The early experimental gardens led to programs in all of the schools under official sanction and funding by the Grand Rapids Board of Education and were also funded by multiple business and land owners. One school in particular became a model for Grand Rapids schools, gaining national attention.

Oakdale School was located in a mostly second-generation Dutch enclave and had a ninety-by-ninety-foot garden plot dug into the school grounds during the summer of 1913. This lot was prepared through a partnership between Principal Frances Van Buren and the Association of Commerce (precursor to the Chamber of Commerce). Half was a community garden where neighbors grew flowers that the Oakdale Mothers Club delivered to the local hospital. The other half was split into garden plots for individual students. About a quarter of the students were also cultivating home gardens. Many of the vegetables were canned with the help of the Oakdale Mothers Club, and the surplus was donated to the neighborhood children's home (*GRP* 1 October 1912, 9; 7 July 1914, 1; 8 July 1914, 6; Van Buren 1913).

Van Buren's work gave the Grand Rapids school gardening programs national visibility. She had forged a strong relationship with Louise Klein Miller, curator of school gardens in the Cleveland school system where Superintendent Elson had transferred in 1906 (*School Journal* 1 April 1905, 347–50). Miller promoted school gardens as a tool to fix "defective children," clean up poor sections of the city, and lower the cost of living (*GRP* 7 October 1913, 2). Van Buren's professional connections led to presentations at national conventions and articles written for academic and popular journals reporting on the successes of the Oakdale gardens.

National visibility increased when the Grand Rapids public schools enlisted the help of E. C. Lindeman of the Michigan Agricultural College for school gardens during the summer of 1915. Lindeman published an instructional manual based on a program called the Grand Rapids Home Garden Plan for School Children, which was distributed to city leaders across the country. The Grand Rapids Plan brought together a coalition of multiple actors including the Board of Education, Kent County Farm Bureau, Michigan Agricultural College Extension, US Department of Agriculture, and local bankers (Lindeman 1916; *GRP* 18 September 1916, 21; *Social Service Review* December 1916, 15–17). Other supporters included newly established neighborhood improvement district boards and the Social Welfare Society.[3]

As the United States entered the First World War, Grand Rapids became a leader in urban agriculture. According to some reports, over five thousand acres split into twenty-eight thousand lots were under food production and were being worked by over thirty thousand families in the summer of 1917. The Victory Garden movement had more financial backing, media exposure, and general goodwill than any previous gardening movement in the city. The Board of Education retained a full-time city-wide garden supervisor, and the Old National Bank, with the help of the local Kent County agricultural specialist and the Association of Commerce, secured two agents from the Michigan Agricultural College who initiated canning cooperatives, organized vacant

Figure 2. The Grand Rapids Home Garden Plan for School Children was a collaborative effort by public and private organizations focusing on school and backyard food production. It was the culmination of a decade of work by women's clubs to get children in Grand Rapids interested in gardening. This pamphlet, published in 1916, gives an overview of the 1915 growing season and was distributed to city leaders and gardening advocates across the country. Image courtesy of the Bentley Historical Library at the University of Michigan, with special thanks to Lindsay Smith.

lot gardening, and worked with neighborhoods on community gardens (GRH 22 July 1917, 4).

The city government was also busy streamlining connections between landowners and potential gardeners. Mayor George P. Tilma had a plan written whereby a record would be kept of "all worthy citizens in humble circumstances, particularly city dependents, those receiving aid from churches, benevolent and charitable societies and all others who may be worthy, though not strictly dependent, to classify same, but under no circumstances whatever to divulge their identity" (Tilma 1917). The council felt that Tilma's plan would waste time finding "indigents" that "don't care a rap" about growing food. Instead, they approved an ordinance whereby the comptroller's office would simply answer calls from an advertisement in the *Grand Rapids Herald* asking for "worthy people." Over three hundred lots were apportioned on the first day (GRH 26 March 1917, 3, 20 March 1917, 1; GRP 27 March 1917, 9).

Mayor Tilma was in office for only one year on account of a rewriting of the City Charter, which put a manager-commissioner system in place and ultimately shifted electoral power from working-class neighborhoods (Kleiman 2006). However, his interest in improving the well-being of working families in Grand Rapids led to lasting changes. Six months prior to the vacant-lot gardening program, he had been working with the home economics department of the Grand Rapids Federation of Women's Clubs (GRFWC) to find ways to lower the cost of living in Grand Rapids. The GRFWC was a strong supporter of domestic sciences in the local schools and was made up of many of the same women from the Ladies Literary Club. Their meetings resulted in the mayor-appointed High Cost of Living Commission, which was responsible for the creation of the three public retail farmers markets.

TILMA'S HIGH COST OF LIVING COMMISSION AND FARMERS MARKETS

The High Cost of Living Commission, which was formed in 1916, had four females among its eleven inaugural members. At least two of these women were also part of the Civic Health and Beauty Committee of the Ladies Literary Club. It should be noted that these women were appointed to the High Cost of Living Commission prior to the ratification of the Nineteenth Amendment. It is very possible that this was the first time women were part of a mayor-appointed commission in Grand Rapids. Gaining a female perspective, however, was earning favor across the United States. As the Nobel Prize–winning suffragist Jane Addams had claimed a decade earlier, "City housekeeping has failed partly because women, the traditional housekeepers, have not been consulted as to its multiform activities" (Addams 1978). The commission's job was to investigate the price of milk and the feasibility of a retail market in Grand Rapids, filling in the gaps left after the Grand Rapids Federation of Women's Clubs had conducted its own local food inquiry.

While women's clubs in larger cities like New York, Chicago, and Pittsburgh promoted city-wide boycotts of food retailers who were accused of unnecessarily high prices, the GRFWC visited with grocers and food dealers to uncover causes of high food prices (*Augusta Chronicle* 29 November 1916, 2; GRH 7 December 1916, 1). Their inquiry did not find blame among the local grocers for high prices; rather, the grocers convinced them that middlemen were liable. In fact, club women invited all women in the city to a meeting where they publicly absolved the grocers and wholesale dealers, but not the milk dealers, who were not willing to share their books with the women. Mayor Tilma had met privately with the federation's Home Economics Committee prior to the public meeting to discuss the high cost of living and mentioned his support for public retail farmers markets (GRH 13 December 1916, 1; Grand Rapids Federation of Women's Clubs 1916). After a couple of mass meetings where all of Grand Rapids was invited to discuss high prices, the mayor put his new commission to work. The first action by this commission was to find a space for a retail farmers market.

Eva Hamilton was appointed chairperson for the High Cost of Living Commission's Market Site Committee. Hamilton researched the success of markets in similar-sized cities and used this information to overcome resistance to retail markets in the City Council. She brokered a deal with the owner of land on the west side of the river, gaining rent-free access for the first experimental season (Hamilton 1917). The Leonard Market opened on July 7, 1917, with a speech by newly elected mayor Philo C. Fuller, who voiced his full support for the market and asked all citizens to cooperate in any way to fight the high cost of living (*Grand Rapids News* 7 July 1917).

Figure 3. The Leonard Street Farmers Market was the first successful public retail market in Grand Rapids, Michigan. Pictured here in 1927, it opened in the summer of 1917 and was improved and expanded over the next decade. Located in a Polish and Lithuanian neighborhood, it was the result of work by women's organizations looking to lower living costs for working families. Photo courtesy of the Grand Rapids City Archives, with special thanks to William Cunningham.

Fuller was the first Grand Rapids mayor under a new manager-commission political structure that greatly weakened the voting power of the west-side neighborhoods of Grand Rapids. The Leonard Street Market was placed in a Polish, Lithuanian, and Dutch west-side neighborhood sharply opposed to the new charter. Indeed, this neighborhood housed the participants of the furniture strike in 1911, which was the impetus for government restructuring. The two other markets, the South Side Market and East Side Market (renamed the Fulton Street Market in 1923), were also placed in immigrant neighborhoods, which were hostile to the charter change. Like the first school gardens, the markets were explicitly focused on providing food for working-class families. However, the newly organized city commission's support for the market may indicate an attempt to pacify ill feelings toward the government structure. More research would be needed to uncover hidden motivations for the location of these markets.

One of the other original members of the High Cost of Living Commission, Emily Chamberlin, continued the crusade for retail markets as Hamilton refocused her efforts toward suffrage and state politics. Chamberlin sent letters to the council as chairperson of both the Food Production and Marketing Committee of the Women's Committee of the Council of National Defense and the Civics Committee of the General Federation of Women's Clubs. Her letters were accompanied by petitions signed by hundreds of housewives in the south end of the city (Chamberlin 1918). This led to the opening of the South Side Market in 1918. The Fulton Street Market opened in 1922 following a similar petition drive.

An amendment was made to the city charter through a popular vote, which gave the city of Grand Rapids authorization for purchasing retail market sites through the Public Service Department, and the Leonard Street Market site was officially purchased for $15,000. The South Side Market and the Fulton Street Market sites were also purchased by the city, improved through municipal funding, and managed by city employees. By 1923, the markets had a combined average of almost ten thousand people each day of operation, with a total attendance of over six hundred thousand between May and December (Grand Rapids Public Service Department 1924). According to reports on the markets, customers could find better prices and fresher food at the markets if they were willing to cultivate relationships with the farmers, and the smaller farmers were getting a better return on their products. Also, an observed unintended effect of the market was lower prices at the surrounding grocers (GRP 10 August 1917, 8; GRP 8 August 1921, 17; GRP 17 September 1921, 13).

Still, prices were sometimes higher at the retail markets during peak season on account of demand for the freshest fruits and vegetables, suggesting that the markets were becoming public spaces that did more than simply provide cheap food. All three markets operated for over fifty years under the direction of city employees. The Fulton Street Market is still in operation and stands as a testament to the work of female leaders in Grand Rapids and as a socially embedded public service and space.

REIMAGINING FARMERS MARKETS AND URBAN AGRICULTURE AS PUBLIC SERVICES

The Fulton Street Farmers Market is the last original public retail market in Grand Rapids that operates in its primary location. Its designation as a public service, however, has become less clear through its transition to its own nonprofit organization. The market has a new organizational structure following a multimillion-dollar facelift realized primarily through private funds that were leveraged through a feasibility study that proved the economic benefits of the market. This paradigm shift from public service to economic engine may threaten the sustainability and civic nature of food system activities that have improved urban food biomes of the United States. It tends to be relatively simple, for example, to leverage private financial support for an upscale marketplace in a downtown area that will be the anchor of revitalization, or to find funding for a young entrepreneur looking to cultivate empty lots to supply restaurateurs. Nevertheless, support for markets and gardening programs that simply provide healthy food for people in underserved neighborhoods can be difficult. I believe this is in part because of the ignorance of the historical role such programs have played in our cities.

The feasibility study for the Fulton Street Farmers Market was conducted the season following my three-year tenure as market manager, during which the market had successfully transferred interim management to the Midtown Neighborhood Association. The out-of-town consultants' lack of historical knowledge of a midsized city farmers market was immediately apparent and was reflected in the opening paragraph of their report: "Historically, markets typically evolved over time from open-air to a market shed to a market hall, and perhaps finally, to a market district" (O'Neil et al. 2008). This may have been the case in large cities like Detroit and Cleveland. However, the Fulton Street Farmers Market had ninety years to evolve, and it has not drastically changed since its inception in 1922. Its existence, as shown above, was actually in opposition to the new market district created in 1897 and was not meant to be a destination to attract business outside its neighborhood.

Of course, I am not suggesting that creative urban food systems should not be economically viable. However, I can attest from my own experience as manager of one of the busiest farmers markets in Michigan that turning a significant profit would be difficult. The market can sustain itself but is by no means a successful business. The retail markets in Grand Rapids have never been able to recover their development costs without outside help. This is why they have traditionally been public services provided through the municipality.

Only when farmers markets and urban agriculture are once again considered necessary public services will we have vibrant marketplaces and urban gardening programs where they are most needed. A clearer understanding of local history can help facilitate these undertakings, as the origins of our local food organizations can quickly be forgotten. Dory Thrasher found this in her participation with the Brooklyn Community Garden, which amplified already tense racial and economic relations partly because it became disconnected from the history of the community. Knowing where we come from can not only help avoid such mistakes but can also inspire us when we are faced with opposition. Learning about the struggles of historical figures such as Eva Hamilton empowers us to question local policy and influence decisions that will improve the well-being of those who live and eat in the cities of the United States.

NOTES

1. "Public retail market" is the historical term for what we now consider a farmers market. "Farmers market" could apply to the many types of marketplaces in early-twentieth-century cities, including wholesale markets, terminal markets, and municipal markets. I use the terms retail market, public retail market, and farmers market interchangeably to refer to grower-only, retail farmers markets (Brown 2001).
2. Municipal governments were very active in promoting various fresh produce marketing schemes during the early part of the twentieth century. These included street peddling, curb markets, and other alternatives. Existing research has focused primarily on large cities such as Chicago and New York (Morales 2000; Donofrio 2007).
3. The Social Welfare Society attempted a vacant lot gardening program in 1894 based on the "potato patch" model followed by Mayor Hazen Pingree in Detroit to help alleviate the problems of unemployment

after the Scare of 1893. Only a handful of articles and scattered meeting minutes could be found on this movement in Grand Rapids.

DISCUSSION QUESTIONS

1. What factors were behind the forcing of vegetable farmers from the wealthier and higher-priced fruit vendors? Why may the desire for separation be so severe as to request "police coercion"?
2. Otto offers many criticisms of the moving of the market in 1897. Might there have been a legitimate market reason for moving the market? How democratic was the initial decision on the market move?
3. Consider various food vendors today. Do we see anything resembling the conflict between commercial and small-scale food vendors in Grand Rapids at the turn of the century?
4. The author raised the issue that the school gardens in Grand Rapids served more of a moral and educational purpose than a nutritional one. Was the Health and Beauty Committee's contest in the same vein except on a larger scale, or did it ultimately serve a different function?
5. Consider the woman who expressed incredulity toward the wealthy Eva Hamilton "tell[ing] the poor how they should save." Was this the grumbling of one person in Grand Rapids, or does it speak to a larger issue of paternalism in urban agriculture that lingers today?
6. Were individual actors like Eva Hamilton and Mayor Tilma the primary instruments of change, or was it social and economic movements in general?
7. The significant role of women in this case study contributed greatly to the relative success of urban agriculture in Grand Rapids. Do you think opposition to more equitable urban gardening is still tied to gender or race?
9. The Fulton Street Market was a success, as it survives and thrives today. What factors contributed to its survival where other markets have failed?

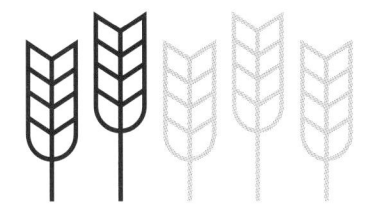

SECTION TWO
Regulation

Different approaches to regulation depend on the organizations, personalities, geography, economy, environment, and politics of existing legal structures. Social movement can motivate new policy (Covert and Morales 2014) but might also produce piecemeal policy. However, when cities and regional entities see the benefit of using urban agriculture and related food systems to integrate across separate domains, then they can find reasons to collaborate and achieve mutually advantageous goals. Of course there will be conflicts; cultivating compost or livestock may provide educational opportunities as well as employment, but will these be popular with neighbors? In the absence of recreational opportunities, will community gardens be the best use of open space? Chapters in this section provide us with examples of these opportunities and tensions as well as the means to understand and apply their insights.

CHAPTER 4

Urban Ag' in the 'Burbs

MEGAN HORST, CATHERINE BRINKLEY,
AND KARA MARTIN

INTRODUCTION

In this chapter, we call for increased attention to *suburban*-agriculture policy. In the past decade, great strides have been made to readapt appropriate urban agriculture in many urban areas. Most of this attention has been centered in large cities. Suburbs have not been the focus of reforms or of academic investigation. However, suburbs represent a significant and mostly untapped potential space for agricultural policy. Suburbs contain the majority of the US population and urban land. Suburban agriculture also has a history of significantly supporting food-stressed countries (Brinkley 2013).

Suburban agriculture will benefit from a different policy approach than urban agriculture. Urban and suburban agriculture are different in their histories, demographics, land-use mixes, plot dimensions, and preexisting prohibitive zoning regulations. As such, suburban planners and policy makers do not find the practices of larger cities particularly relevant to their community context. Yet without other resources, practitioners continue to draw from urban-agriculture models in large cities. City "best practices" regarding allowed uses, lot size, guidelines on aesthetics, monitoring and permitting, and restrictions and requirements for additional buildings, animals, fences, parking, setbacks, lighting and noise, odor, and waste management (for examples, see chapter 16) often need to be changed to suit the suburban context. Overall, we see a need for more attention to suburbs, and purposeful inclusion of diverse jurisdictions in research and development of best practices.

In this chapter, through a series of four cases, we explore modern suburban agriculture to identify patterns of early adoptions and limits. We conclude with three takeaways regarding urban agriculture, as well as a set of recommended steps for suburban-agriculture policy making.

HISTORY OF URBAN AGRICULTURE

To understand the current regulations governing urban agriculture, it is helpful to know how urban agriculture became so heavily regulated, and often discouraged, across the United States. This was not always the case. Among early colonial American cities, urban agriculture was emphasized as a key component in civic life and the food supply (Vitiello and Brinkley 2013). Philadelphia, Baltimore, and Boston maintained planned commons for animal pasturing, communal gardens, and designated areas for private gardening. Spanish settlements in the West were constructed following planning documents that specified the distance of resident housing to farmland and gardens to make food transport efficient. However, urban agriculture fell out of favor in the late 1800s as inventions like refrigerated railcars enabled the transport of food over long distances, removing the necessity of relying on nearby production (Cronon 1991).

Significant changes occurred regarding policies around animals. Early American cities relied on animal agriculture for waste management, transportation, and food supply (Brinkley and Vitiello 2013). Free-roaming hogs cleaned up household slop and processed swill in what would become New York's Central Park before the city had a public suburban sanitation department. Sheep and goats grazed on the Boston Commons. Chickens and roosters squawked as they scratched through the dirt. Animals were everywhere, as was their manure.

As cities grew and modernized, concern also grew over animals as nuisances. Specifically, there were fears over the spread of cholera from swill-feeding pig operations, and a desire to relocate the urban poor and the waste services they provided (via pigs processing slop and urban dairy cattle processing spent grains from distilleries, for example) to less-central locations like the suburbs. Cities started government sanitation services at the turn of the twentieth century, outlawing piggeries, composting, and informal garbage collection (for more on this issue, see chapter 6). Soon thereafter, city planners, veterinarians, and newly created public health officials engaged in efforts to remove animals from cities. Prohibitions on animals were codified in zoning ordinances in the 1920s and still exist today, almost a century later.

In addition to concerns over animals, some (though not all) early city officials also considered inner-city food gardening problematic. While early planners saw gardens on the residential fringes (early suburbs) as important amenities that could sometimes be used for cultivating vegetables and fruit, they debated why and if the inner-city poor should garden. At one of the first city planning conferences, planner Thomas Adams noted that urban agriculture was prevalent in poorer quarters of the city; when "you go down into these slums . . . you will find people trying to grow potatoes in a box one foot square . . . and these people will tell you that they are living under these conditions only because they cannot do otherwise" (*Proceedings of the Third National Conference* 1911). Public health concerns over the quality of food, the beauty of gardens, the rats and other pests, and disease potential provided a basis for regulating urban agriculture out of the poorer quarters of the city and preemptively banning it from newly created wealthier suburbs.

Enforcement of these regulations was uneven and incomplete, and many people conducted urban agriculture without legal reprimand. Even today, immigrants and low-income communities in cities across the country continue to keep chickens, goats, potbellied pigs, and other animals in the city, in some cases illegally (Brinkley and Vitiello 2013).

Early concerns over aesthetics, values, and public health continue to serve today as challenges to reintroducing agriculture in both cities and suburbs. Planners struggle with both real and imagined health concerns, and with improving poorer neighborhoods without disenfranchising their residents. However, the benefits of urban agriculture are increasingly recognized (Stocker and Barnett 1998; Martin and Marsden 1999). Environmental engineers (Hynes 1996), planners (Martin and Marsden 1999), and landscape architects (Lawson 2005; Helphand 2006) have argued that urban agriculture promotes community and economic development by beautifying neighborhoods, fostering food security and healthy eating, engendering higher property values, promoting civic engagement, providing cost savings and income generation opportunities, improving mental health, reducing the urban heat island effect, and promoting water filtration (Brinkley 2013; Viljoen and Howe 2005). Chapters 2, 3, 14, and 16 in this book also discuss some of the benefits of urban agriculture. These benefits likely hold true for suburban agriculture as well.

With the renewed focus on urban agriculture, some planners, local governments, designers, and developers are reintegrating food practices that were zoned out a century ago. Cities from Cleveland to San Diego have liberalized their regulations for agriculture (Mukherji and Morales 2010; and as discussed in detail in chapter 13 in this volume). Many new municipal sustainability plans promote urban farms and composting (see chapter 6 for more on this issue), and a few big cities allow households to keep goats and potbellied pigs. Suburban agriculture is taking off in some suburbs, too. Notably, development-supported agriculture, in which a developer includes some form of food production—a farm, community garden, orchard, livestock operation, edible park—in planned housing developments, is growing around the country. There are more than two hundred agriculturally based developments nationwide (Runyon 2013). Meanwhile, some established suburbs are identifying opportunities to reintegrate agriculture into their suburban fabric.

DIFFERENCES BETWEEN URBAN AND SUBURBAN AGRICULTURE

It is important to note that not all suburbs are the same. There is a range of socioeconomic characteristics of their populations, regional relationships, land-use patterns and forms, social norms, leadership and staff interest and capacity, and existing policy infrastructure. For example, suburbs vary in both size and population. Some have high rates of diversity and others are more homogeneous in terms of income and ethnicity. Inner-ring suburbs often have downtowns and possess mixed-use neighborhoods, a grid pattern, high levels of walkability, and access to transit. Exurban communities tend to have segregated land uses and curving subdivisions, looping streets, and cul-de-sacs. These differences indicate that planning for suburban agriculture is very contextual and political.

With these differences in mind, we note that suburbs also face different contexts than cities in terms of US policy. To date, most of the academic literature and professional publications have focused on urban agriculture in cities but not suburbs. Suburban agriculture (emphasis intended) differs from urban agriculture in many respects, including the community context, planning context, and, likely, form and activities.

First, the community context is different in suburbs in three main ways: socioeconomic characteristics of the population, land use and social norms, and community interest. Regarding socioeconomic characteristics, over half of Americans live in suburbs. Both poverty and immigration are growing faster in suburbs than in cities in the largest metropolitan areas (Kneebone and Berube 2013). As a result, many suburban areas with little experience with either immigration or poverty face new public policy challenges. Some suburbs also have higher rates of chronic diseases, such as childhood and adult obesity, hypertension, and heart disease, than urban areas (Lake and Townshend 2006). It appears that these characteristics, combined with others such as lack of walkability, mean that suburban agriculture may be just as tightly linked, if not more so, to healthy food access and issues of social justice as agriculture in urban areas. Other important aspects of the suburban community context are land use and social norms. For example, suburbs tend to have larger lot sizes and lower population density. One example of how these differences play out is in setting minimum and maximum lot sizes for urban agriculture. For example, the city of Seattle allows farms of up to four thousand square feet before plan review is required, which means that most residential properties would never need plan review. In suburbs around Seattle where lot size is much higher, that may be a limiting maximum. There are also different social standards for landscaping, with suburban areas sometimes valuing manicured or grassy lawns over food production.

Another major difference between suburban and urban areas is the existing planning context. Planned communities with a homeowners association, common in suburbs across the United States, often have layout designs, planning, and bylaws that preclude residents from food gardening (UrbanAgLaw.org, 2013). The local government often has very little influence over these regulations. Staffing and resources may also be more limited in suburbs. While cities like New York and Seattle have recently hired food system planners, suburbs often operate with much smaller planning departments and budgets and lack the capacity to have staff dedicated to urban-agriculture issues. This has implications for how agriculture gets institutionalized in each jurisdiction. Big cities may be able to take on programming, for example, a city-managed community gardening program. Suburbs with fewer resources may need to engage in cost-neutral activities, like enabling urban agriculture by local churches, nonprofit organizations, neighborhood groups, and schools.

Finally, the form, uses, and activities of agriculture may differ in suburban areas. In cities, particularly in dense downtowns, space is often more limited and expensive. Urban agriculture may take the form of container or rooftop gardening. Community gardens likely consist of a large number of plots packed densely onto a small site where urban agriculture is often conducted for small-scale production—such as for personal use by apartment dwellers or restaurants for growing herbs. In less-dense suburbs, however, agriculturists may be able to access larger plots of land. At the same time, larger-scale operations may be more likely to come in conflict with suburban neighbors over concerns about noise (from chickens, goats, and tractors, for example), dust, smells, and aesthetics. This again has implications for policy, as larger operations might be more accepted if, for example, they were set back from neighboring property lines or shielded with a hedge, a row of trees, or a fence.

EXAMPLES: SUBURBAN-AGRICULTURE POLICY CHANGE

We now present a series of brief vignettes intended to illuminate the unfolding story of modern suburban-agriculture policy. We highlight four suburban cities from across the United States that have made recent policy changes regarding gardens, animal husbandry, and/or on-site sales. The municipalities were selected via a snowball sampling technique, using our personal connections and e-mail listservs. They are intended to represent a snapshot of evolving suburban-agriculture policy, rather than best practices.

The case suburbs exhibit fairly traditional suburban form, with some diversity in their specific characteristics and demographics (table 1). In choosing these cases, we explicitly did not consider rural towns or new developments (so-called farming subdivisions or agrihoods) that integrate farming operations with residential developments, such as Prairie Crossing, Illinois; Agritopia, Arizona; or the Bucking Horse development in Fort Collins, Colorado (Cohen 2013; Murphy 2014). We also intentionally did not focus on urban cities, though some of their efforts may have relevance for suburbs. In particular, the city of Toronto recently proposed a new Residential Apartment Commercial Zone, which allows for-profit gardening in apartment tower clusters. We also did not look at county-level planning for urban agriculture, since counties typically do not have authority over regulations in their cities (though this varies). However, many counties are fostering urban agriculture, and this has an influence on suburbs within their jurisdiction. Examples are Salt Lake County, Utah, which has implemented an Urban Farming Program to lease unused public land to farmers, and Arlington County, Virginia, which recently established an Urban Agriculture Task Force and intends to adopt a Food Action Plan. Meanwhile, Prince George's County, a suburban county in Maryland, recently passed legislation to allow urban farms in multifamily zones.

In focusing on suburban cities, our search uncovered wide variation in interest and initiative in suburban-agriculture policy. On the one hand, during our search for cases we heard anecdotally that many suburbs are not explicitly engaging in in-depth or innovative agricultural policy. For example, in Montgomery County, Maryland, despite a concerted effort to examine agricultural policies, "nothing coalesced and no policy was adopted. Not so much for a lack of interest but rather a perception that current policy was doing enough and other things demanded finite attention." Similarly, in New Jersey, it was noted that "there are lots of examples; however, most of them in suburban areas are only small steps toward urban agriculture." Finally, in Washington's Puget Sound region, results of a scan of food policies in over sixty cities revealed that while the large urban cities of Seattle and Tacoma each have over a dozen policies supportive of urban agriculture in their comprehensive plans, very few suburbs have explicit policies regarding suburban agriculture in their comprehensive plans or codes (Horst 2012). In fact, many existing policies are unclear or unsupportive and may even serve as barriers. As an example, numerous cities have code restrictions against vegetation over one foot high, vegetable beds or fruit trees in the front yard, or composting, and some require significant fencing and lighting for gardens.

While seemingly not the norm, some suburbs are implementing changes to their suburban-agriculture policy. Four examples are discussed below, with attention paid to the context and process, the source of support and resistance, the relationship between community organizations and local government, specific strategies of change, obstacles or barriers, and the impacts and future opportunities. For each example, we interviewed a city planner directly involved with the change and/or a community leader active in advocacy efforts.

Royal Oak, Michigan

Royal Oak was founded in the late 1800s, and its historic downtown has recently attracted new development. The city's population is predominantly white, middle class, and family oriented. The city is surrounded on all sides by other urban, developed areas and has no large tracts of farmland. In 2011, the city of Royal Oak revised its zoning ordinance to define and support home gardens, community gardens, and for-market farms. Prior to this, nothing in the zoning ordinance had explicitly allowed or enabled agriculture. Such activities were not listed as either a permitted or special land use in any zoning district. An organization looking to establish a community farm on a vacant lot adjacent to a school approached the city looking for direction and advice. As a city planner noted, "we had nothing specific to offer, no way to respond." The petitioners subsequently sought and received a use variance from the Zoning Board of Appeals. Anticipating future requests, city staff set out to clarify their code. Staff did their homework, looking for examples and best practices. Not finding cases of other suburban cities like Royal Oak, they drew mainly from American Planning Association resources and models from larger cities like Madison, Minneapolis, Austin, and Boston. While Royal Oak was similar to Detroit, Royal Oak staff were not strongly influenced by Detroit's recent policy work on urban agriculture. As the interviewee noted, "that's a lot different scale [sic]."

Ultimately, the city implemented new definitions and standards for domestic, community, and market gardens in its zoning code. As noted, the purpose and intent is to "ensure that domestic, community and market gardens are

Table 1. Quick profiles of case suburbs

	Royal Oak, MI	Wheat Ridge, CO	Snoqualmie, WA	Federal Way, WA
Population, 2012 estimate	58,410	30,717	11,591	91,923
Type of suburb	Inner ring with historic downtown	Inner ring with agricultural history	Outer ring (exurb) residential with historic downtown and agricultural history	Outer ring residential
Median income, 2007–2011	$38,205	$48,777	$120,714	$55,846

Source: US Census Bureau.

appropriately located and protected to meet the need and demand for local food production, and to enhance community health, community education, garden-related job training, natural resource protection, preservation of green space, and community enjoyment. Because they will typically exist in close proximity to residential uses, concern will be given to ensuring compatibility between uses." The new changes provide enhanced clarification and flexibility, as well as fairly strict regulation. Regarding the former, the new changes explicitly allow the three types of gardens in most kinds of zoning. They also permit direct on-site sales at seasonal farm stands. Regarding strict regulation, the changes establish some constraints. For example, domestic gardens are explicitly not allowed in a front yard. Meanwhile, community and market gardens are required to be screened from view with a wall, fencing, and/or landscaping at least six feet in height along all side and rear lot lines. Finally, most composting (including of fruits, vegetables, and animal waste) is prohibited in all garden types. This suburb continues to have century-old concerns over agriculture's compatibility with existing uses. It is moving cautiously to enable suburban agriculture but wants to keep it mostly out of sight. Meanwhile, tight composting regulations, reflecting concerns about rodents and odors, prevent its link with waste management.

A planner at Royal Oak noted that the new changes have "not had much of an impact." There has not been any obvious increase in the amount of gardening, and in fact, the proposed garden that initially set the changes in motion has been put on hold as the lot was developed. The City Commission has begun initial talks around other aspects of urban agriculture, but this remains to be played out.

Wheat Ridge, Colorado

Wheat Ridge was a thriving agricultural community in the mid-1900s, consisting of mid- to large-scale orchards and crop fields. Partly in response and resistance to urban growth, the city turned down annexation opportunities and incorporated in 1969. Today, Wheat Ridge is an inner-ring suburb of Denver, Colorado, and is home to approximately thirty-one thousand residents. Its population is older, with a smaller average household size and slightly lower median income than in neighboring communities.

The policy approach by local government to suburban agriculture has roots in the area's agricultural history. Current residents still identify strongly with a farming identity. For example, the city has long had (in the words of a local planner) "very permissive" ordinances related to large animals, poultry, and bees. However, residents and organizations recently sought clarity on other agricultural practices. One source of advocacy on these issues was Live Well Wheat Ridge, one of twenty such coalitions in Colorado. Live Well Wheat Ridge, including collaborators from local businesses and nonprofits as well as the police and public health departments, advocates for local healthy eating and active living policies and infrastructure improvements. Live Well Wheat Ridge conducted a community needs assessment in the fall of 2010, identifying significant interest by residents in locally grown foods. Live Well Wheat Ridge also organized community advocacy to call for and support new suburban-agriculture legislation. One representative noted that Live Well Wheat Ridge "provided the why, marshaled the community and engaged elected officials." A big part of their advocacy was emphasizing the multiple benefits of urban agriculture to the local economy and health.

The city noted the need to update suburban-agriculture regulations during the process of adopting the Comprehensive Plan in 2009. Subsequently, in 2010, planning staff initiated a process to change the zoning code. They conducted five formal meetings in addition to other, softer forms of community outreach like casual conversations. Staff drew from examples in cities from inside and outside Colorado, including Boulder, Chicago, Cleveland, Minneapolis, Milwaukee, and Seattle. The City Council unanimously adopted Ordinance 1491 in 2011. One change is that community gardens (under the category of "urban gardens"), farmers markets, and produce stands are now allowed in any zone district. Meanwhile, under homeowner occupancy regulations, residents are allowed to grow and sell produce and keep animals. The city also eased permitting requirements for hoop houses.

Overall, the city opted for a political approach, supported by the mayor at the time, of "least restrictive environment." The changes enable residents and organizations to engage in agricultural activities without creating additional demands on the city. The community has received the results of the policy changes warmly, and city staff have not heard any complaints. As one planner noted, "no news is good news on that end." The policy changes appear to have enabled on-the-ground action. For example, several entrepreneurs have established commercial urban farms and initiated community harvest swaps. A participant from Live Well Wheat Ridge hopes that the momentum and goodwill built during the suburban-agriculture policy process can extend forward into other, related healthy community planning initiatives, such as active transportation.

King County, Washington

In spring 2010, Public Health–Seattle and King County was awarded a federal stimulus grant to address rising obesity rates in the region as part of the Centers for Disease Control and Prevention's Communities Putting Prevention to Work (CPPW) project. As part of this project, six suburban cities received subawards to assess and implement local policies to improve food access and opportunities for physical activity for their residents. This included considering suburban-agriculture policies as a means for residents to grow their own healthy, fresh foods. The King County suburban cities of Snoqualmie and Federal Way, both described below, have since adopted changes in suburban-agriculture codes and policies.

Snoqualmie, Washington   Snoqualmie (pop. 11,591) is a small rural exurb on the outskirts of King County, in the foothills of the Cascade Range. It has seen significant population growth in the past decade. Once the biggest hop producer in the state, the city now produces food mainly in residents' home gardens and chicken coops and two community gardens. The city's Park Department manages two community gardens. In addition, the Ridge, a planned mixed-use housing development in which a majority of the population resides, has a community garden for residents. However, the Ridge's homeowners association currently prohibits front-yard food gardens.

The suburban-agriculture policies adopted in Snoqualmie have focused primarily on community gardening. This was largely because of interest in the city's P-Patch program. To ensure the program's expansion along with future population growth, the city adopted a level of service standard (i.e., the number of garden plots available per capita), along with policies to support youth involvement in food growing and preparation and prioritizing community garden development in underserved areas. While there was no outward opposition from the community or elected officials, there was also little active support in the policy development and adoption process. The policy effort was accomplished largely by a dedicated planner, who prioritized the work and led the necessary cross-departmental coordination, with staff time and technical assistance funded by CPPW.

The community garden policies were adopted into the 2012 Open Space, Parks and Recreation Plan, seen as the most appropriate place for adoption. However, food production or broader food access does not necessarily fall neatly into the parks' main focus of providing recreation opportunities. The planner has reflected that in hindsight, the policies may have been better placed in one of the Comprehensive Plan's elements (e.g., Community Character or Environment). City staff may explore this organization in future policy updates.

Federal Way, Washington   Federal Way (pop. 91,923) is the third-largest suburb in King County and has significantly high obesity rates (more than

28 percent of adults are obese). Prior to receiving the CPPW award, the city of Federal Way had little experience with suburban-agriculture policy making aside from a 2011 update on its animal code to allow for chickens and ducks, but not roosters. During the CPPW project (2010–2012), city planning staff became familiar with broader food access issues in the city. At the same time, staff received inquiries from the community regarding where food production was allowed and to what extent. A second opportunity for technical assistance was made available through a grant-funded partnership with a local nonprofit, Forterra, which provided urban-agriculture policy expertise and community outreach support. Without outside resources such as CPPW and Forterra, staff would have been limited in collecting stakeholder input, and the policy development itself would not have been as in-depth.

In January 2013, goals and policies were adopted into the Comprehensive Plan's Land Use element that directed staff to establish a new zoning ordinance around suburban agriculture. Throughout the policy development and adoption process, there was positive support from the various stakeholders. Planning commissioners and City Council members attended community events held by Forterra to solicit input on their interests and concerns. In addition, residents attended commission and council meetings to give public comment in support of the proposed policies. Discussions among the planning department, planning commission, City Council, and residents focused on the nuances of the policy and codes themselves. Questions came up such as, what is appropriate for residential versus nonresidential areas? How prescriptive should the city be? What is the city's capacity to regulate and monitor?

The city turned toward examples of model policies and codes from larger, denser cities and found them useful in overall concept. However, use of these "big city" codes required some adaptation through either modifications such as lot size requirements or simplifications of the code, which, if left as it was, would have required regulatory monitoring the city did not have the resources for. At times, some code was not even considered, such as regulations for rooftop gardening, currently irrelevant in the suburban context. By year's end, the City Council had unanimously approved the urban-agriculture ordinance that included definitions (i.e., urban agriculture, community garden, urban farm, farmers market, cottage food operation, farm stand), specified the

Table 2. Brief summary of changes in urban-agriculture policy

	Comprehensive Plan or other plan changes	Code and regulatory changes
Royal Oak, MI		Enhanced definitions, clarity, and some restrictions. Allowed three types of gardens in most kinds of zoning; permitted direct on-site sales at seasonal farm stands; and established constraints on front-yard gardening and composting as well as fencing/screening requirements.
Wheat Ridge, CO		Established least restrictive environment: community gardens, farmers markets, and produce stands allowed in any zone district; residents allowed to grow and sell produce and keep animals; eased permitting requirements for hoop houses.
Snoqualmie, WA	Established level of service standard for community gardens in Open Space, Parks and Recreation Plan.	
Federal Way, WA	Included goals and policies to establish new zoning ordinance. Included urban agriculture in Comprehensive Plan.	Established definitions, zones, and operation standards.

zones in which each use is allowed, and listed operation standards (e.g., signage, hours of operations).

The changes implemented by each suburban city regarding urban-agriculture policy are summarized in table 2.

CONCLUSIONS

This exploration into suburban-agriculture policy begins with the observation that resident demand and interest by elected officials and individual staff (so-called policy entrepreneurs) are important drivers of change. In the cases considered, the process was initiated via some expression of community interest (a theme also discussed in chapters 3 and 13). As a participant in this research put it, "folks with dirt under their fingernails are the ones making those connections rather than the policy-makers." In addition, individual staff played instrumental roles in moving the wheels of bureaucracy. They also needed support from those with ultimate decision-making authority, such as the City Council. The motivations for government staff to support suburban agriculture ranged from concern about healthy food access, to viewing food as part of a broader transition to sustainability, to passion for developing community self-resilience. Other motivators for suburban-agriculture policy change included public health departments, which provided some funding and technical skills. This suggests that education and training for policy makers, planners, and public health officials may be an effective way to increase suburban-agriculture policy interest.

We end with three main takeaways for future suburban-agriculture policy making. One is that community norms play a big role. Across the board, community and for-profit gardens seem widely accepted and make for easier policy change. However, activities like chicken raising, composting, and front-yard gardening may be acceptable in some suburbs and some populations, and controversial in others. In some suburbs, historical concerns about urban agriculture, such as its negative impact on property values, remain alive and well today. If supportive suburban-agriculture policy is to be implemented more widely in suburbs, these lingering resistances must be addressed and a new kind of suburban dream or aesthetic better appreciated.

A second takeaway is that suburban agriculture may be a stepping stone toward more coordinated and broader food systems planning in suburbs. For many suburban planning and public health departments, suburban-agriculture policy represents a low hanging fruit, where other options to encourage healthy, local food systems (such as regulating fast-food stores or protecting farmland) may not be currently politically feasible. Meanwhile, policy changes to support suburban agriculture do not necessarily entail a great commitment of resources; rather, policy change can be used to facilitate or at least permit actions by other actors at low or no cost to the city. As such, suburban-agriculture policy may represent a form of pragmatic incrementalism toward broader food systems and healthy community planning.

Finally, we highlight the need for further suburban-agriculture policy research, for example on interest in suburban agriculture among immigrant communities; on the trade-offs associated with restrictions and requirements about fences, setbacks, signs, and noise; and on the impacts of policy change on measurements of local food access and health. We conclude by offering a short set of action steps for suburban communities that are interested in fostering suburban agriculture.

1. Articulate a clear goal of fostering suburban agriculture (including community, private, and market gardens, animal raising, small-scale processing, and direct on-site suburban agriculture) in local policy and planning. Connect this to other local planning goals such as promoting health, well-being, social justice, economic development, and community identity.
2. Identify existing programs, organizations, and key individuals that either do or could engage with suburban agriculture, including potential partners such as the health department and agricultural extension offices.
3. Create multiple and welcoming opportunities for the above partners, as well as residents (particularly from low-income, immigrant, and other groups with food access and food security challenges) and other stakeholders to participate in the planning process.
4. Collaborate with partners to conduct an audit of local policies,

including those in specific areas of the city's Comprehensive Plan, such as Economic Development, Open Space and Parks, Health, and Sustainability plans, and municipal code and development regulations. Identify policies that are completely missing, need clarification, present potential barriers to the practice or expansion of suburban agriculture, or can be expanded to include explicit support for and regulation of suburban agriculture. A policy audit could focus on suburban agriculture holistically (as in the case of the 2012 project in Milwaukee, Wisconsin) or on a targeted subject, such as community gardens.
5. Develop appropriate responses by drawing from and adapting the policies in available tool kits and resources, including a 2011 American Planning Association report on urban agriculture. Consider adding supportive policies regarding animals, on-site sales, and medium and small-scale community, private, and for-profit agriculture to the Comprehensive Plan and other topical plans; clarifying and permitting urban-agriculture land uses and activities in the city code; reducing restrictions and requirements on additional buildings, animals, fences, permitting, parking, setbacks, lighting, and the management of waste, noise, and odor; using city funds and resources to support programming on community gardening and urban agriculture; encouraging businesses, programs, and uses that support local food production; identifying public lands suitable for a demonstration project or community garden expansion; and promoting alternative growing options, like school and roof gardens, orchards, and edible landscaping.
6. Evaluate impacts and check in with key stakeholders about lingering concerns, gaps, and barriers, and modify accordingly.
7. Share experiences with other suburban communities.

ACKNOWLEDGMENTS

We wish to acknowledge and thank our interview participants, including those for cases not used in this chapter. We also thank Eva Ringstrom for her contributions. Any omissions or errors in the synthesis are ours.

DISCUSSION QUESTIONS

1. Is the separation of livestock from cities as a mainstream governmental practice a positive or negative development for society?
2. What can urban planners and city officials do to address legitimate health concerns about suburban agriculture, public health, and aesthetics while enabling agriculture for its other benefits?
3. Does the current lack of suburban agriculture stem from de facto bans, a lack of will, or something else entirely? Explain and elaborate.
4. Imagine that suburban agriculture has flourished to its fullest extent. How does this change urban agriculture and conventional large-scale agricultural practices and the food systems shared among urbanized areas?
5. Wheat Ridge, Colorado, experienced some degree of success in enabling suburban agriculture. What components might allow suburban communities to replicate the success of this suburb?
6. The authors note that suburban communities are seeing an overall rise in poverty and health concerns. How might a rise in suburban agriculture in these communities combat these trends?
7. A common feature among the case studies is that the three suburban communities borrowed from larger cities in crafting agricultural policies (as opposed to borrowing from rural agricultural policies). Why might suburban communities look to cities for policies instead of rural areas? How might their resulting agricultural policies differ if they borrowed from rural areas?

CHAPTER 5

Cultivating in Cascadia

Urban-Agriculture Policy and Practice in
Portland, Seattle, and Vancouver

NATHAN MCCLINTOCK AND MICHAEL SIMPSON

INTRODUCTION

For decades, proponents of "Cascadia" have called for an independent political entity that would incorporate the US Pacific Northwest states of Oregon and Washington and the Canadian West Coast province of British Columbia. While political secession remains largely a fantasy, there are nevertheless aspects of a shared history and geography that have informed a regional identity. The Cascade Range and its lush forests of Douglas firs unify the physical geography across all three jurisdictions. A common history of Euro-American colonization a century and a half ago and a subsequent economy centered on forestry and resource extraction have also contributed to the region's unity. Much more recently, the values of environmental sustainability and livability have added a new dimension to this common identity.

Urban agriculture figures prominently in Cascadia's cityscapes. It serves as part of a shared cultural identity for many residents and is embraced by local governments in Cascadia's three major cities, Portland, Seattle, and Vancouver. While its contribution to local food supplies is modest (and incredibly difficult to quantify), urban agriculture nevertheless figures centrally in regional visions of a sustainable food system. The goal of this chapter is to examine urban agriculture's similarities and differences in these three cities in Cascadia, with particular attention to the role of planning, policy, and municipal support. Although our primary focus is on how urban agriculture is practiced, we begin with a brief overview of official urban-agriculture policy and programs in the three cities that draws from municipal policy documents as well as interviews with practitioners and policy makers conducted between October 2012 and May 2013. We then present the results of a 2013 survey of businesses and organizations that practice urban agriculture in Cascadia's three major cities. These findings offer a glimpse of the landscape and character of urban agriculture as it is practiced on the ground today.

The region's three major metropolitan centers—Portland, Seattle, and Vancouver—all wear this mantle of sustainability with pride. All have topped various sustainability and livability rankings over the past decade, garnering worldwide attention and accolades for their advances in green infrastructure, from light rail and bike lanes to green buildings and gray-water recycling. The region's reputation as a mecca of livability, however, has translated into a steep rise in the cost of living in all three cities. Median home values in Vancouver are higher than anywhere else in North America, and Seattle's are not that far behind. Although property values in Portland are not as elevated as in these other two cities, they still remain high relative to household income, in part because of the city's urban growth boundary and emphasis on increasing density in the urban core.

Of course, for every commonality that the region shares, there are at least as many differences. Different economies, changing demographics, separate regulatory frameworks, unique social histories, and an international border have together shaped Portland, Seattle, and Vancouver into quite distinct cities, belying the notion of regional uniformity. For instance, while all three municipalities have similar populations, the population of the Seattle metropolitan area is significantly larger than that of either Portland or Vancouver (see table 3). Vancouver, while smaller than Seattle, is Canada's third-largest city and therefore plays a more important role nationally; home to the country's largest

Table 3. Demographic and socioeconomic indicators

Indicator	Portland	Seattle	Vancouver
Population[a]	583,776 (2010) 594,687 (2014)	610,298 (2010) 624,681 (2014)	603,502 (2011) 635,660 (2014)
Population density[a]	4,457 / sq mi	7,441 / sq mi	13,590 / sq mi
Metro population[a]	2.23M (2010) 2.31M (2014)	3.45M (2010) 3.61M (2014)	2.37M (2011) 2.47M (2014)
Median home value[b]	USD $284,900	USD $433,800	CAD $752,016
Median household income[b]	USD $52,657	USD $65,277	CAD $68,970
Population in poverty (%)[b]	12.0	7.2	20.8[c]
White (%)[b]	72.0	66.7	46.7
Black (%)[b]	6.1	7.2	1.0
Latino (%)[b]	9.4	6.4	1.6
Asian (%)[b]	7.4	14.0	47.1
Native American / First Nations (%)[b]	0.6	0.5	2.0
Other (%)[b]	0.7	0.7	1.7
Multiracial (%)[b]	3.7	4.6	1.5

[a] 2014 population estimates from American Community Survey 5-Year Estimates (2010–2014) and Statistics Canada projections (2014); 2010 US Decennial Census and 2011 Census of Canada (2011).

[b] American Community Survey 5-Year Estimates (2009–2013) and 2011 Census of Canada.

[c] Since Statistics Canada does not calculate a poverty threshold, we report the low-income cutoff (LICO) statistic here. It therefore cannot be compared to US poverty rates.

port and a high immigrant population, Vancouver serves as Canada's western gateway to the world. It also has nearly two to three times the population density of the other two Cascadian metropolises. Like Vancouver, Seattle is a global city. Home to aerospace and high-tech industries, as well as major ports, it is the de facto economic capital of the Pacific Northwest. Historian Carl Abbott (2001, 40) describes Portland, on the other hand, as "a regional city in contrast to networked Seattle," one that historically looked inward to domestic markets. More recently, however, Portland has capitalized on its ability "to bring environmentalism and urbanism together in a coherent package of mutually supportive planning and development decisions" (6) by exporting its expertise in sustainability planning and green economic development around the world.

THE SUSTAINABILITY SHIFT: SCALING UP URBAN AGRICULTURE THROUGH POLICY AND PLANNING

Interest in urban agriculture in Cascadia's cities has historically tended to parallel the ebbs and flows of national and regional economic drivers, much as it has in cities throughout North America (for instance, see chapters 2 and 3 in this volume). The current resurgence of community gardening and urban agriculture in Cascadia is often traced to the early 1970s, when the energy crisis, rising costs of food, and a vocal countercultural movement embracing environmentalist ideals each contributed to a do-it-yourself, "back to the land" ethic (Sanders 2010; Hou, Johnson, and Lawson 2009). Despite the upsurge of grassroots interest, however, support for urban agriculture from municipalities remained lukewarm for several decades. Only in the late 1990s and the early 2000s did municipal support for urban agriculture begin to shift in Portland, Seattle, and Vancouver. At that time, the concept of sustainability, first formalized by the United Nations Brundtland Commission in 1987 and the Rio Conference in 1992, began to infuse municipal discourse in the region. Growing municipal commitment to urban agriculture followed. In the words of Portland's first Community Gardens manager, Leslie Pohl, "It took thirty years until I felt like [Portland's Community Gardens program] was pretty secure"; only once the city began to take an interest in issues of sustainability in the early 2000s did the Community Gardens program get "more respect"

(Pohl 2013). This more recent support for community gardening after decades of uncertainty seems to mirror a pattern seen elsewhere across the United States, as reported in chapter 13 on New York City's GreenThumb program.

In Portland, local policy and planning efforts have begun to link food production with larger environmental priorities. Portland's *Climate Action Plan* in 2009 was the city's third plan aiming to reduce carbon emissions, but only the first to identify "Food and Agriculture" as a central component of these efforts; at the county level, the 2010 *Multnomah Food Action Plan* has called for a "sustainable food system" by 2025. The city of Portland has supported the acquisition of several larger properties for the purposes of farm and garden education, and it doubled the number of community garden plots in the city between 2009 and 2012, adding over one thousand new plots in seventeen new gardens under its One Thousand Gardens Initiative. It also partnered with a local nonprofit to establish four new community orchards, and in 2012 it introduced sweeping changes to the zoning code intended to facilitate neighborhood-scale food production and distribution (see tables 4 and 5).

In Seattle, urban-agriculture efforts since the 1970s have focused primarily on the P-Patch community gardening program. More recently, however, the city has demonstrated its commitment to a broader vision of urban agriculture. City officials declared 2010 the Year of Urban Agriculture. That same year, the city changed its zoning code to increase the limit on domestic fowl from three to eight, provide a height exception to rooftop greenhouses, and allow community gardens in all zones, among other adjustments aimed at encouraging greater production. Recent planning and policy documents have also affirmed support for local food system and urban-agriculture initiatives. For instance, Seattle's *Local Food Action Initiative* called for the city to create a policy framework that would "strengthen the region's food system in a sustainable and secure way" (City of Seattle 2008), and the recent *Food Action Plan* (City of Seattle 2012) places local food production as one of four goals for the food system (see tables 4 and 5).

In Vancouver, the City Council adopted a definition of sustainability and a set of related principles in 2002 to guide the municipal government's actions and operations. The following year, it applied this new sustainability lens to the food system by approving a motion calling for the creation of a "just and sustainable food system" and adopting the *Action Plan for Creating a Just and*

Sustainable Food System. When former organic farmer Gregor Robertson was elected mayor in 2008, the food system took on an even more important role in municipal sustainability efforts. City officials vowed to add 2,010 community gardening plots before hosting the 2010 Winter Olympic Games, a benchmark that the city successfully exceeded. Soon after, Vancouver's *Greenest City Action Plan* proclaimed its goal of becoming the world's "greenest city." The 2011 document further pledged that the city would be a "global leader in urban food systems" (City of Vancouver 2011). Over the past decade, Vancouver has legalized backyard beekeeping and chicken raising and, most recently, adopted the *Vancouver Food Strategy*, which addresses the entire food system under a unified policy umbrella (see tables 4 and 5).

In sum, urban agriculture in Portland, Seattle, and Vancouver has shared a common trajectory since the 1970s. Garden projects that began as idealistic community-driven endeavors were later adopted into official city community gardening programs. More recently, urban agriculture has been incorporated into broader sustainability and food systems policy frameworks. Of course, municipal policy and programs are only one of the many social, economic, and political forces shaping Cascadia's urban-agriculture landscape. While government programs and policies certainly influence the way people practice urban agriculture in each city, such local initiatives have often been after-the-fact responses to the advocacy and innovations of community activists and organizations in the city. Regardless of what municipalities in Cascadia think urban agriculture *should* look like, every urban agriculturist is motivated by his or her own values and perspectives; similarly, each navigates local policy in his or her unique—and often quite creative—fashion. We now turn our attention to these motivations and practices.

THE LANDSCAPE OF URBAN-AGRICULTURE PRACTICE

Types of urban-agriculture organizations

The landscape of urban-agriculture practitioners in Cascadia today is vast; not only are urban-agriculture businesses and organizations numerous, but they engage in diverse activities (see fig. 4). In March 2013 we conducted a survey of North American urban-agriculture organizations and businesses. Among our responses, we heard from forty-nine organizations and twen-

Table 4. Scale and type of urban gardening

Garden type	Portland[a]	Seattle[b]	Vancouver[c]
Community gardens	50 (2,100 plots)	81 (2,650 plots)	97 (3,900 plots)
Year Community Gardens program was established	1975	1973	1995
Management of Community Gardens program	Portland Parks and Recreation	Department of Neighborhoods	Vancouver Park Board
Community orchards	4	37	3
Commercial market gardens and urban farms	32	2 municipally run; private unknown	17 (as of 2011)
Edible school gardens (public schools)	80	33[d]	50–60[e]

Sources:
[a]Portland Department of Parks and Recreation; Portland Bureau of Planning and Sustainability.
[b]Seattle Office of Sustainability and Environment.
[c]City of Vancouver.
[d]Washington State Department of Agriculture Farm to School (2011).
[e]Estimate, Vancouver Coastal Health (2013).

ty-two businesses practicing urban agriculture in the Portland (n = 43), Seattle (n = 12), and Vancouver (n = 16) regions. Our survey respondents included privately owned businesses, nonprofit organizations, grassroots collectives, government institutions, cooperatives, schools, colleges, student groups, private clubs, and religious organizations. In what follows, "businesses"

Table 5. Key plans and policies related to urban agriculture in Portland, Seattle, and Vancouver

Level or sector of policy intervention	Portland	Seattle	Vancouver
Community gardens	----	----	Park Board Community Gardening Policy (1995) Greenest City Action Plan (2011)
Vacant land	Diggable City (2005)	An Inventory of Public Lands Suitable for Community Gardening (2008)	City of Vancouver Urban Agriculture Inventory (2006)
Animal husbandry	Title 13 of City Code	City Code Update (2010)	Animal control bylaw and zoning; development land use zoning bylaw
Beekeeping	Title 13 of City Code	City Code Update (2010)	Health and safety bylaw
Zoning code	Urban Food Zoning Code Update (2011)	City Code Update (2010)	Urban Agriculture Guidelines for the Private Realm (2009)
Climate	Climate Action Plan (2009)	Climate Action Plan (2013) Climate Adaptation Strategy (2013)	Community Climate Action Plan (2005) Climate Change Adaptation Strategy (2012)
Food system	Multnomah Food Action Plan (2010)	Seattle Food Action Plan (2013)	Vancouver Food Charter (2007) Vancouver Food Strategy (2013)

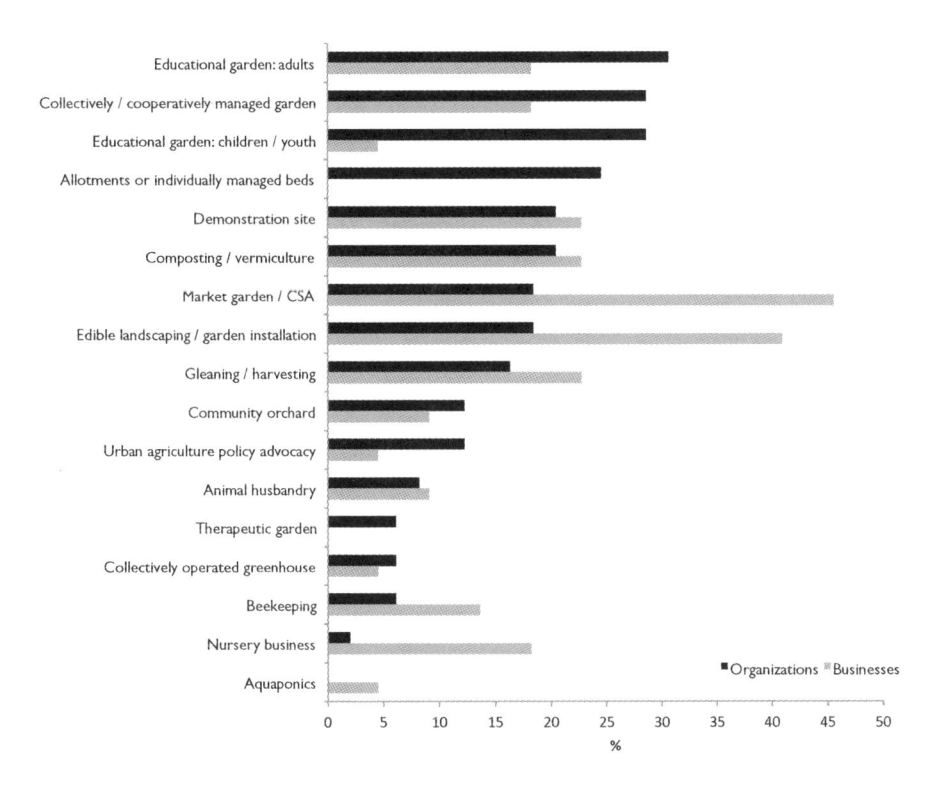

Figure 4. Primary types of urban agriculture practiced by organizations and businesses.

means privately owned businesses, and "organizations" means all other types of organizations.

Most of the respondents—organizations and businesses alike—operate market gardens or community-supported agriculture (CSA) programs (an alternative distribution model where customers generally pay at the beginning of the season for a weekly produce box). Most of the surveyed businesses earn revenue by installing gardens or edible landscaping. Organizations, on the other hand, more frequently run educational, allotment, and collective gardens. Other common practices among businesses and organizations alike include gleaning and harvesting, composting, and operating nurseries and demonstration sites. Many also practice composting and beekeeping and engage in urban-agriculture policy advocacy. For many organizations, such activities are secondary. More than 40 percent of businesses also engage in urban-agriculture policy advocacy, while just under 40 percent of organizations engage in educational activities for children and youth. Composting was shown to be another very common practice among urban agriculturists, with over half of the respondents indicating that they engaged in this practice as either a primary focus or a secondary activity, a finding that supports Suerth's observations in this volume about decentralized composting (see chapter 6).

Motivations of practitioners

What motivates these businesses and organizations is as wide-ranging as the types of agriculture they practice (fig. 5). In their responses, many urban agriculturists underscored the multiple public benefits that urban agriculture provides. The most common motivations for engaging in urban agriculture as stated by businesses and organizations very much reflect the progressive environmental and social concerns commonly associated with the region (see fig. 5). Among our respondents, more than 60 percent of businesses and organizations ranked sustainability and environmental or agroecological concerns as primary motivations. This tendency of Cascadia's organizations to link urban agriculture with environmental concerns and with the larger food system is consistent with the discourse employed in municipal planning and policy documents over the past decade, as discussed above.

Possible future scenarios of ecological and economic collapse further motivate some urban agriculturists. One Vancouver market gardener we interviewed explained, "When there is a water shortage in California, another water shortage in California, a trucking strike, fuel prices go through the roof because there's another war and we need to grow food in the city like we've done in the past, we've got a model to build on." Many emphasized "stacked functions"—a term frequently invoked in permaculture design to suggest that any single element can have multiple beneficial functions in an ecological system—to describe the multiple focal areas of their project, highlighting primary motivations such as community building, education, food security, public health, fresh food, and sustainability, among others.

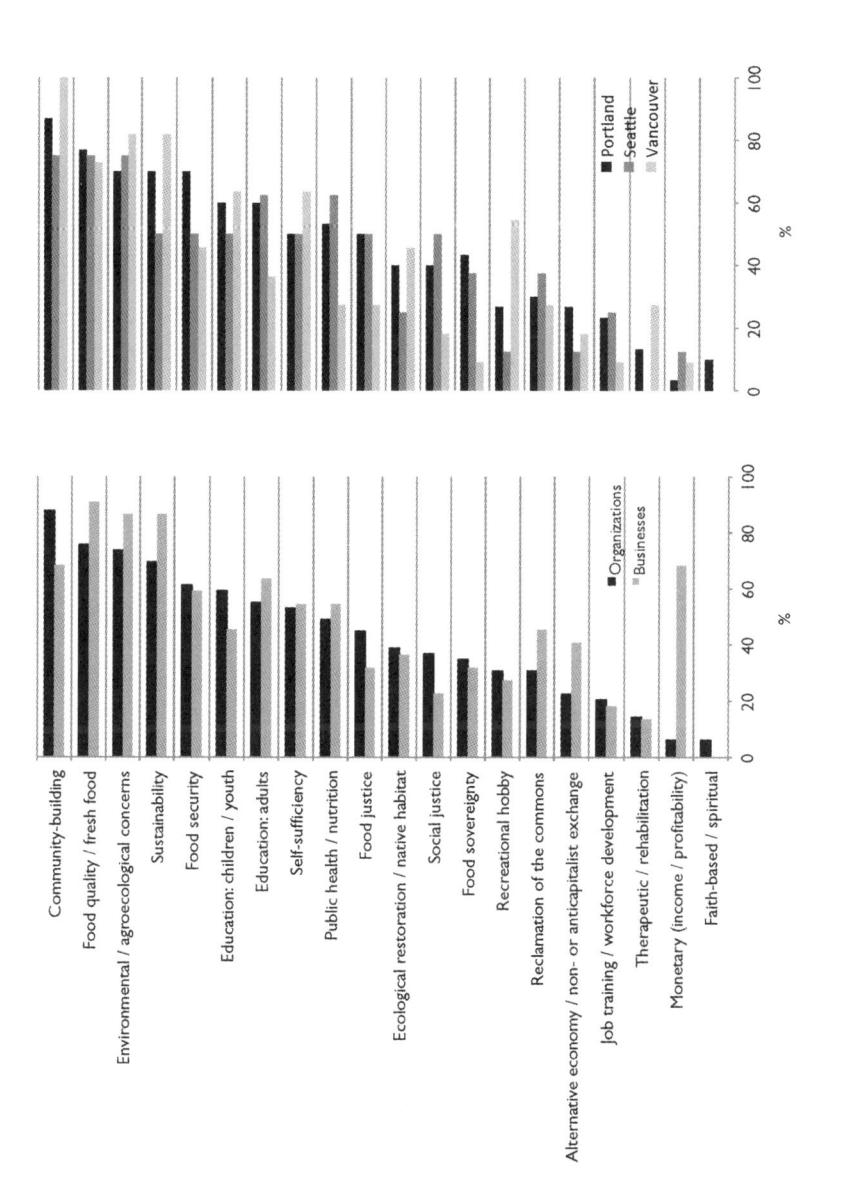

Figure 5. Motivations of organizations and businesses (left) and motivations of organizations disaggregated by city (right).

Many organizations are motivated by concern over community food security. Discourse on community food security took hold in the 1990s as activists began to draw attention to issues of inequitable access to food in low-income communities (Gottlieb and Joshi 2010). Survey results reveal a strong emphasis on food security. One Portland organization responds to "the need to grow our own food and raise our own animals to provide for our families and friends who need to be fed. Food is too expensive to buy." Similarly, a Seattle organization member responded, "We believe in the power of edible neighborhoods to end hunger and increase health in our communities. We get there by supporting people in building and growing gardens." Most are also motivated by food quality or freshness. One Portland urban farm provides "fresh organic local vegetables for neighbors to support a vibrant local community and economy, provide healthy food, and reduce the distance that food travels from field to plate. We also seek to establish the farm as an educational and community hub for the neighborhood." Nearly all organizations in the three cities emphasized community building as a primary motivation, or, in the words of one Portland organization member, "to feed ourselves where we live and build more rooted community in the process." Finally, one Seattle urban agriculturist is motivated by a "desire to make better use of neglected public space."

Interestingly, however, very few organizations indicated that monetary income or profitability is a primary motivation. Indeed, many practitioners share the belief that urban agriculture is not a huge moneymaker. As an urban-agriculture business owner in Portland affirmed, "There is no money in urban agriculture. Forget health insurance or any savings." Similarly, a Vancouver market gardener told us,

> Economically you're dealing with something that isn't necessarily so stable, right? . . . The fact is that you're living in Vancouver and look at how expensive it is to live here. And you know, you need to be supported by your partner that's doing whatever. One guy was saying he was getting paid twenty-three cents an hour, when all was said and done. I'm sure he ate really well, but it's going to be a big strain on anybody doing it to make a business work. It's such a new thing for [new urban farmers], because seeing what you need to produce to make money in agriculture, and not be subsidized and not have big grants, is insane.

While many of those surveyed are simply not motivated by profit, some go a step further and suggest that their endeavors are actually anticapitalist or a means of creating an alternative economy. A member of one Portland organization expressed a desire to operate outside the dominant market logic of capitalism:

> In reclaiming our ability to grow our own food . . . we are able to act independently of large-scale corporations and capitalistic business models that actively destroy intact ecologies in the name of profit. Along these lines, we are able to shift power back into the hands of ourselves and our neighbors through the creation of communal systems for food production and distribution, thereby increasing our health, reducing any negative impacts on our local environment, and gathering together under a common purpose and identity.

Moreover, although monetary income and profitability are understandably a primary motivation for more businesses than organizations, it is interesting to note that more than 40 percent of the surveyed businesses also identified "reclamation of the commons" and "alternative economy / non- or anticapitalist exchange" as primary motivations, with no notable variation between cities (see fig. 5). One Portland business owner stated, "We are alarmed by the extent of our societal and planetary crisis, but even more motivated by the creative possibilities for renewal and re-emergence." A Vancouver business owner was motivated by the "ability for growing to have positive presence in low income communities and improve quality of life." A Seattle business owner wrote, "Building skills and economy within neighborhoods builds community, increases food security, adds to a biodiverse urban landscape, creates eddies of social life, and overall is about resiliency and solutions." He notes that these ideas are incorporated in the company's charter. Responses such as these suggest that some urban-agriculture activists in Cascadia may be using the legal entity of a business strategically in order to pursue broader objectives of systemic transformation.

Indeed, the tendency of practitioners to emphasize urban agriculture's multiple functions while downplaying its economic benefits is consistent with what we heard from policy makers in the three cities. As the Vancouver Food Policy Council cochair Brent Mansfield expressed, "Maybe the economic development argument is not enough. The proof is in the pudding . . . we don't see it yet. There's not many commercially viable urban farms, right? It's not a motivation." Food policy advisers in all three cities confirmed that when they advocate for public investment in urban agriculture, it is generally more persuasive to emphasize urban agriculture's multiple benefits than it is to make a purely economic case. Seattle's food policy adviser told us, "I find the economic argument hard to make. If there is some compelling reason why I think urban agriculture needs to happen in a space where there's some competing demands that are of a higher value . . . I would be more inclined to fall back on all of these multiple goals rather than to say economically it actually is a better thing, because there often isn't the evidence for that."

Similarly, Portland's Food Policy and Programs manager, Steve Cohen, stated, "I'm not really sure that we could make a strong case that [urban agriculture] was going to change economics. But, for recreation, for community capacity, I think that that's really strong for us in terms of getting together and doing it at the neighborhood level, it's really the most important one for us."

Farming the margins

So why is it so difficult to make a case for urban agriculture's economic viability? High property values and relatively low returns on produce are the primary obstacles to profit-generating models of urban agriculture in Cascadia's three metropolitan areas. As one Portland urban farmer with a year-to-year lease affirmed, "Probably the biggest issue that I have is land tenure, because this is really becoming a more and more valuable piece of property. A New Season's [upscale supermarket] was just built, you know, and they're breaking ground across the street from New Season's for another condo . . . I would love to be able to keep this [land], but there's no way that I can."

Indeed, according to market logic, crop production is a far cry from the "highest and best use" of market-rate land. Rising land values in the three cities, particularly in once-dilapidated areas that are now undergoing redevelopment, further threaten urban agriculture. As noted by the Vancouver Food Policy Council cochair, "I don't feel confident, with land values the way they are, to say, 'Yes, this economically is worthwhile.' . . . In Vancouver, with housing prices, [urban agriculture]'s never gonna come out as the winner" (Mansfield

2013). These concerns about the economic viability of urban agriculture in cities with high property values were not lost on earlier generations of urban agriculturists, as described in chapter 2.

Given the high market value of land in Portland, Seattle, and Vancouver, large-scale production is rare. Within the city limits, urban-agriculture projects are typically relegated to very small parcels of land, or in some cases rooftops, balconies, and marginal spaces with limited development potential. In Seattle and Vancouver, more than a third of respondents produce on less than 2,500 square feet, compared to lower-density Portland, where only 18 percent farm on such small parcels. Nearly half (48 percent) of all respondents indicated that their businesses or organizations produce in front yards, side yards, or backyards. Perhaps because of the city's exceedingly high property values, respondents from Vancouver singled out access to land as a major need for organizations.

Given the "market illogic" of urban food production, most of the businesses and organizations that we surveyed operate at the margin on shoestring budgets. About 20 percent of urban-agriculture programs surveyed operate on budgets of $1,000 or less, and an additional 25 percent operate with budgets between $1,000 and $5,000. The businesses that we surveyed have budgets that are generally larger than those of urban-agriculture organizations, but their budgets are also small. More than half (55 percent) of surveyed businesses allocate $10,000 or less to their urban-agriculture activities; two-thirds allocate $50,000 or less. Given the small operating budgets and high cost of land, urban agriculture in Cascadia, as in other places with high land values, requires subsidies of one form or another, including labor, land, funding, and other forms of governmental support. We will discuss these requirements in turn.

Volunteer labor

One way that urban-agriculture projects—organizations, in particular—are able to resourcefully operate on small budgets is by minimizing the number of paid staff that they employ, relying instead on volunteer labor. Twenty-two of forty-nine organizations surveyed reported having no paid employees working on urban agriculture. Thirty of forty-nine reported that they rely "very much" or "tremendously" on volunteer labor or community engagement. About three-quarters of surveyed organizations rely on ten or more volunteers per year, and a fifth of all organizations rely on a hundred or more volunteers annually. While businesses do engage volunteers, they rely on them less than organizations do. Only nine of the twenty-two surveyed businesses reported that they rely "tremendously" on volunteerism. More than half rely on five or fewer volunteers.

Alternative land tenure arrangements

To overcome the high cost of land, urban agriculturists in all three cities have secured a variety of alternative tenure arrangements. While many of the businesses and organizations surveyed operate their farming or gardening activities on private land, relatively few actually own the parcel. While seven of twenty-two businesses reported owning the property that they farm, only seven of forty-nine organizations reported owning the property. Leasing private land was no more common; only eight of twenty-two businesses and five of forty-nine organizations reported leasing private land for production.

Rather, the majority of urban agriculturists acquire access to the land from the property owner for free, for a nominal or symbolic fee, or for some sort of in-kind donation (e.g., of produce, labor, grounds maintenance, or other benefits). Some organizations and businesses (such as Portland's Urban Farm Collective and Vancouver's Inner City Farms) rely on a land tenure model in which they farm the yards of residential landowners. This type of arrangement allows for a more profitable urban-agriculture model because the landowner effectively absorbs the cost of the land. This is particularly important for businesses trying to make a profit. While only 35 percent of organizations produce on residential land, more than 90 percent of the businesses surveyed cultivate on residential property (two-thirds of this on single-family residential lots).

Other urban agriculturists rely on free access to public land. More than a third of surveyed businesses (36 percent) and organizations (37 percent) alike produce in public parks or on land belonging to other public institutions. Similarly, schools and universities provide land to a significant proportion of both organizations (41 percent) and businesses (23 percent). Production on *vacant* public land, however, tends to be more common for organizations (29 percent) than for businesses (9 percent). The type of arrangement for access to public land varies; in some cases, the arrangement is stipulated through a binding lease, while in other cases, there is a more informal agreement. Among organizations, nearly half of those surveyed (twenty-four of forty-nine) noted that they are granted access to public land for free under the terms of a formal agreement. Among businesses, seven of the twenty-two surveyed operate on public land using a lease agreement, while seven others have no formal lease. Three businesses (two in Portland and one in Seattle) reported using public land without permission. Only one organization reported using public land without permission.

While public land may be accessed for free, accessing it is not always the easiest path to take. Providing organizations or businesses with access to public land can be politically challenging if the general public believes that public land is being used for private gain. To alleviate this risk, one model being tested in Seattle is a lease agreement that grants public land access to a nonprofit organization at the market rate. Instead of paying this rate monetarily, however, the group agrees to provide an equivalent value in public benefit through their work (Lerman 2013).

Grant funding

In addition to relying on access to free or reduced-cost land and free labor, most urban-agriculture organizations also depend on external sources of support for their projects. While surveyed businesses earn, on average, 90 percent of their revenues from sales, organizations depend heavily on grants, both from government agencies (46 percent of revenues) and from private foundations (31 percent of revenues). Reliance on government grants appears to vary between cities; on average, government funding makes up, on average, about 72 percent of the budget of surveyed organizations in Vancouver, compared to 40 percent among Portland organizations and 55 percent among Seattle organizations. The sources of these grants also differ. Only a few organizations in Seattle and Vancouver reported receiving municipal grants for urban-agriculture projects, and none reported receiving grants from county or regional governments. For Portland organizations, on the other hand, municipal and county or regional grants were the most common forms of governmental support. These differing results may reflect the fact that over the past few years, several urban-agriculture organizations in Portland have received funding through the Portland Development Commission's Community Livability grants program and the East Multnomah Soil and Water Conservation District's Partners in Conservation and Small Project and Community Events grants programs.

Other governmental support

In addition to direct funding, urban agriculturists benefit from additional forms of governmental support (see fig. 6). Survey results point to a clear differentiation between cities in terms of the types of support that organizations receive. For instance, the most common support that organizations in Vancouver benefited from was access to water, either for free or at a reduced rate. While organizations in Portland commonly benefited in this way as well, only one organization in Seattle reported receiving free or reduced-fee access to water from the city.

Overall, however, businesses benefited less from governmental support than did organizations. Few of the twenty-two urban-agriculture businesses we surveyed reported significant governmental support for their activities. Of the Vancouver businesses, one received a tax break or other incentive and another received fast-tracked permitting. One business in each city reported receiving access to land for free or at a reduced rate, while one in Portland and one in Seattle received water for free or at a reduced rate. Five cited marketing opportunities, and one each in Seattle and Portland cited training and extension support.

Attitudes toward policy

As discussed earlier in the chapter, urban-agriculture groups have enjoyed a supportive political climate in all three cities over the past few years. As the monetary and nonmonetary forms of support from local government cited above indicate, municipal policy makers have made strides to foster urban food production in a variety of ways, from providing material support to easing restrictions put in place during previous eras. While all three cities have eased restrictions on urban agriculture in bylaws and the zoning code in recent years, not every organization has felt the benefit of such support.

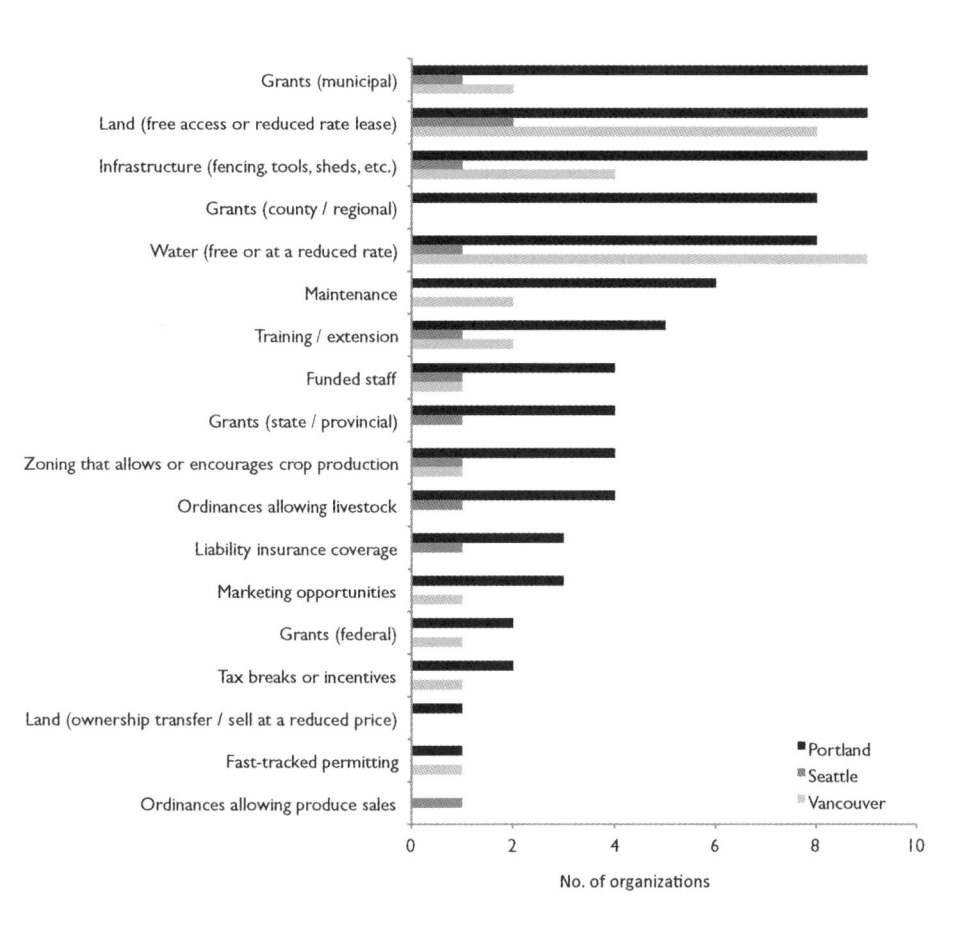

Figure 6. Types of government support received by urban-agriculture organizations.

None of the organizations surveyed from Vancouver, for example, reported having benefited from either ordinances allowing livestock or ordinances allowing the sale of produce, whereas several organizations in Portland reported feeling supported in this way. Some surveyed businesses acknowledged the benefits of municipal policies such as urban-agriculture ordinances; eight of twenty-two acknowledged the support of ordinances allowing or encouraging crop production, whereas seven noted ordinances allowing sales, and six noted ordinances permitting livestock.

Cultivating in Cascadia   77

Some urban agriculturists have been unable to tap into the tiny wellspring of support because of bureaucratic obstacles. For instance, one Seattle co-op explained that because it is not legally licensed as a nonprofit organization, "it's been difficult for us to obtain a lot of what we need on our shoestring budget, such as insurance, renting a building, etc., as we're not eligible for the help that is given to 501(c)(3)'s." At the same time, its geographic location in a city made it ineligible for state funding for rural agricultural cooperatives.

The sheer volume of red tape is a major hindrance for others. One Portland respondent complained that there is "lots of nearby vacant space owned by the city, but at this point it seems to be illegal to use and unrealistic to delve into the bureaucracy." Another group expressed, "Working with the school district has been a real challenge. They approved our project and then back-pedaled on each element that we tried to implement." The comments of another Portland organization hint at some of the politics possibly undergirding such bureaucratic hurdles:

> The bureaucracy of these organizations has hindered their ability and/or desire to work with potential partner organizations, and occasionally with other governmental agencies. For example, in Portland, many involved in the Multnomah County Urban Agriculture Sub-Committee were pressing for changes to zoning to allow for crop trees in certain areas, and advised the Portland Bureau of Sustainability as to what changes could benefit the city. But the Bureau had no power to implement many of those advised changes; instead, the authority fell to other city planners, or to the Parks Department. They had not been involved in the discussions from the outset, and perhaps for that reason were opposed to the advised changes. Eventually, some of the city planners agreed to certain zoning changes. In the case of the Parks Department, however, deep departmental intransigence seemed to create a view that any advice on potential changes to park regulations was nothing but outside meddling in department affairs.

Similarly, a Seattle organization cited departmental infighting, and the mixed messages that resulted, as a major hindrance. Its project is on "land owned by one jurisdiction and stewarded by another. A third jurisdiction/department is charged with fostering community development (including urban agriculture). These agencies have not always played well together. We have sometimes struggled to understand what each wants of us."

Among the seventy-one organizations and businesses that were surveyed, governmental policies appear to pose more of a problem for businesses than for organizations, and more so in Seattle and Vancouver than in Portland. Three of five Vancouver businesses felt hindered by existing regulations, while eight of eleven organizations did not. One respondent complained, "Urban farming is not legalized in Vancouver, which is not a large hindrance, but I would market via farm-gate sales if I was allowed to." Another revealed, "Due to restrictions on urban farming in Vancouver we are in the process of closing down our original site. Vancouver bylaws do not allow for urban agriculture to exist in a for-profit model according to the development and zoning departments of the city." Similarly, in Seattle, three of eight organizations felt hindered by regulations, and three of four businesses expressed the same. One noted, "Publicly owned land and property may not be leased to for-profit businesses for private gain in Washington State." In Portland, on the other hand, only two of thirteen businesses surveyed, and only six of thirty organizations, felt that their activities had been hindered by regulation. Still, despite the fact that our respondents identified regulatory obstacles, they did not tend to believe that changing local laws would benefit their urban-agriculture efforts. This was especially true of Portland respondents.

From this perspective, the relationship between local governments and those urban agriculturists who cultivate the soil seems tenuous, uncertain, and sometimes inconsistent. On the one hand, local governments have become the more recent champions of urban agriculture, having marched forward with a procession of recent plans, documents, and ordinance changes intended to support practitioners; indeed, our survey suggests that urban-agriculture groups have benefited in various ways from such local governmental support. Yet on the other hand, certain groups seem to have benefited more than others. Indeed, as Cohen and Wijsman suggest in chapter 13, new policies and programs can encourage certain urban-agriculture practices and help increase their acceptance into the mainstream, while concurrently inhibiting other such innovations from flourishing. As a result, the organizational landscape—and the attitudes of urban agriculturists—toward policy and planning initiatives are as diverse as the forms of urban agriculture found dotting the landscapes of the three cities.

CONCLUSION

Over the past four decades, urban agriculture has blossomed in Cascadia, growing from a relatively marginal (and sometimes illegal) countercultural practice without municipal support into a movement that is now widely heralded as evidence of local governments' commitment to sustainability. Municipal support in Seattle, Portland, and Vancouver has grown from tepid adoption of community gardens to the active encouragement of urban agriculture that comprises more than gardens alone. Community orchards, "food forests," home-scale production, school gardens, farm and garden education sites, animal husbandry, backyard beekeeping, and commercial farms now all have a place within municipal boundaries in policy and practice alike. Each of the three cities has developed comprehensive food action plans that lay out progressive policy objectives for urban agriculture and visions for a sustainable local food system, more broadly. Urban agriculture is now firmly entrenched in the sustainability efforts of all three cities.

Given the small survey size, our results cannot be generalized to urban-agriculture organizations and businesses as a whole. Nevertheless, we believe that these preliminary results provide important insights into the practices and motivations of urban agriculturists in Cascadia's three metropolitan centers, as well as rich qualitative data on their motivations and challenges that may resonate with policy makers and practitioners in other parts of North America who are interested in learning not only what it takes for urban agriculture to flourish, but also what obstacles might stand in the way. Indeed, despite the progressive, sustainability-minded orientation of each of the cities, advocates are still struggling to put urban agriculture on a more solid footing in Cascadia.

Results confirm that many different types of organizations practice urban agriculture in the three cities, and that the forms of urban agriculture they practice are vast. Despite the diversity, several common trends emerge. Urban-agriculture businesses and organizations typically operate on small budgets. The added pressure of high land values in these cities makes it difficult to practice urban agriculture as a profitable business. Still, profit is not actually a primary motivation for many organizations; remarkably, the same is true for many businesses. Respondents representing both models of urban agriculture more often underscored the numerous other public benefits—from environmental sustainability to community building and food security—as prime motivators. Some, including businesses, suggested that they are even anticapitalist in their motivation.

Importantly, given the difficulties of generating high revenues by selling produce alone, urban-agriculture businesses and organizations are resourceful, tending to rely on other means to keep their projects operational, such as volunteer labor and creative tenure arrangements allowing them to access land for free or at low cost. Further, many organizations receive governmental support to help sustain their projects. Here we do see some notable distinctions between cities, however, revealing the different nature of governmental support for urban-agriculture efforts and its impact on the landscape of practice in the three cities.

Indeed, while businesses and organizations in each city did recognize various ways that urban-agriculture projects had been aided by government, many indicated that government has hindered their project in some way. These hindrances also varied between cities. Profit-oriented models of farming appeared to face more restrictions in Vancouver, whereas jurisdictional obstacles seemed more common in Portland and Seattle. Ultimately, our results seem to indicate that the track record of municipal support for urban agriculture remains mixed in terms of its utility to practitioners. While this might suggest that municipal policy could be changed to support practitioners more effectively, most practitioners remain skeptical that changes in the law would benefit their efforts.

To conclude, urban agriculture in Portland, Seattle, and Vancouver is as similar as it is different. Its commonalities draw from a common regional landscape and history and, perhaps most importantly, a common commitment to sustainability by local government and civil society alike. But the three cities are also each unique in their own right, with different demographics, economies, and policies, which together have shaped urban agriculture in particular fashions. Even within each city, urban agriculture takes various forms and reaps the benefits of municipal policy and funding opportunities differentially. Documenting these differences and similarities, both between and within cities, in an extensive fashion is but a first step in helping to identify both the strengths and shortcomings of urban-agriculture policy innovations in the region. More intensive examination of policy and practice—through case studies, policy analyses, and ethnographies alike—is needed to identify potential best practices for urban-agriculture policy and planning in the three cities, and to foster urban agriculture's contribution to Cascadia's policy commitments to urban sustainability, more broadly.

DISCUSSION QUESTIONS

1. What sort of effect do the high housing values in the three cities—particularly Vancouver—have on the urban-agriculture movement in Cascadia?
2. There are subtle yet important differences among the histories of urban agriculture in the three cities. What are these differences and what do they say about urban agriculture in each city overall?
3. City governments got very involved in promoting urban agriculture in Seattle, Vancouver, and Portland in the last decade. Why might these city governments have moved to embrace and promote urban agriculture more proactively than other city governments in the country?
4. Urban agriculture in Cascadia comes primarily from an environmental and ecological focus, whereas most other urban-agriculture products in this book come from a social and cultural welfare perspective. Considering this, how does this difference make Cascadian urban agriculture unique?
5. What factors account for this difference in approach to urban agriculture? Would a more environmentalist view on urban agriculture fare well in promoting urban agriculture in other parts of the country?
6. A Portland agricultural business owner stated, "There is no money in urban agriculture." Could there be? Should there be? Why or why not?
7. There is also a particularly political element to urban agriculture in Cascadia. Is this element helpful or detrimental? Both? Explain.
8. What issues in urban agriculture are particular to these three cities? Think of the cases in other chapters.
9. Reflect on the land tenure issues and possible solutions presented in the chapter. Do any solutions strike you as more helpful and realistic than the others?

CHAPTER 6

Urban Agriculture

Composting

LAUREN SUERTH

Composting is a natural decomposition process that converts organic materials to usable soil, producing a biologically stable, humic substance that is a nutrient-rich soil amendment (Cooperband 2002). It is a valuable waste management method because it provides environmental, economic, and social justice benefits. Diverting organic materials from landfills to compost operations decreases the amount of greenhouse gas emissions in the air and prolongs the life of existing landfills (Harrison and Richard 1992). Compost is a marketable commodity that, when added to soil, improves the chemical, physical, and biological characteristics of the land, which reduces the need for water, fertilizers, and pesticides (Cooperband 2002). Furthermore, participation in composting can build awareness about the full life cycle of food.

The US Environmental Protection Agency (EPA) has collected and reported data on the generation and disposal of municipal solid waste for more than thirty years. Analyzing these data provides valuable insight into the amount and composition of the waste stream, which, in combination with reemerging environmental and social concerns, allows policy makers, urban agriculturists, and other professionals to identify opportunities to manage and use the system in a more sustainable manner. This chapter focuses specifically on promoting composting in cities because, as of 2010, 80.7 percent of the United States population lived in urban areas, and consequently, our waste management issues are more of an urban phenomenon (US Census Bureau 2013). Composting reduces pollution, lessens pressure on landfills, supports urban agriculture, and has the potential to generate revenue through sales of fertilizer.

The EPA's *Facts and Figures for 2012* report demonstrates first and foremost the increasing presence of waste in modern society. Between 1980 and 2012, the amount of solid waste generated per person per day increased from 3.66 to 4.38 pounds. In 2012, Americans generated about 251 million tons of trash and discarded over 136 million tons to the landfill (54 percent), recycled almost 65 million tons (26 percent), and composted around 21 million tons (8 percent) (US Environmental Protection Agency 2014c). These percentages, however, belie the potential of recycling and composting, because when we examine the composition of the municipal solid waste we find that 55 percent is organic materials (i.e., food scraps, paper or paperboard, and yard trimmings), which are recyclable and compostable. This discrepancy identifies the need to increase the presence and use of waste recovery programs.

My thesis developed from this evidence because it indicates that the existing federal, state, and local waste management policies ignore the environmental consequences of disposing of organic materials in landfills. The objective of this chapter is to identify regulatory methods that legitimize composting within cities. To accomplish this, I argue that reconstructing municipal solid waste regimes will establish composting as a sustainable waste management method as part of the reemerging urban-agriculture movement. As chapter 4 indicates, cities initiated sanitary services early in the twentieth century, outlawing composting and other processes. However, if managed and regulated according to contemporary goals, standards, and uses, composting can support urban-agriculture activities and produce a variety of benefits for communities, as discussed in chapters 5, 7, and 16 on the Pacific Northwest, production processes, and public health, respectively. In this chapter, I also discuss these benefits, but my principal piece of advice is to urge communities to experiment and find an approach or approaches to composting that support their goals and interests.

The chapter will provide the basic context of municipal solid waste regulations for organic materials in order to increase understanding of how local governments can integrate composting into urban areas. Part 1 analyzes composting policies according to two methods, centralized and decentralized. Part 2 explains the structure of composting regulations and how federal, state, and local laws influence municipalities. Parts 3 and 4 provide factors to consider when amending a municipal code to integrate composting into the waste management system and as a permitted land use.

THE MUNICIPAL SOLID WASTE CONTEXT

There are approximately 9,800 curbside recycling and 3,120 community composting programs in the United States, so Americans have opportunities to sustainably dispose of organic materials (US Environmental Protection Agency 2013b, 2014b). These programs, however, are not as ubiquitous as traditional disposal systems, and the amount of organic waste recovered varies significantly according to the material. This is evident by comparing the total amount of a material to its percentage in each waste stream. Figure 7 identifies the total amount of municipal solid waste generated by material in 2012 before recovery, and figure 8 illustrates the proportion of waste that Americans dispose of in landfills, recover through recycling and composting, and combust to create energy.

Analyzing the generation and management data for specific materials demonstrates further discrepancies. For example, in 2012 Americans generated 36.4 million tons of food waste (14.5 percent of the waste stream), disposed of 34.7 million tons (95.2 percent), and composted 1.7 million tons (4.8 percent) of the material. Meanwhile, Americans generated 34.0 million tons of yard trimmings (13.5 percent of the waste stream), disposing of 14.4 million tons (42.3 percent) and composting 19.6 million tons (57.7 percent) (US Environmental Protection Agency 2014c). A comparison of the composition of municipal solid waste with how it is managed shows that the United States recovers an unequal amount of its organic waste and disposes most of the remaining material in a landfill.

This is problematic because when organic material ends up in a landfill, it decomposes and converts to methane, which is a greenhouse gas twenty-five times

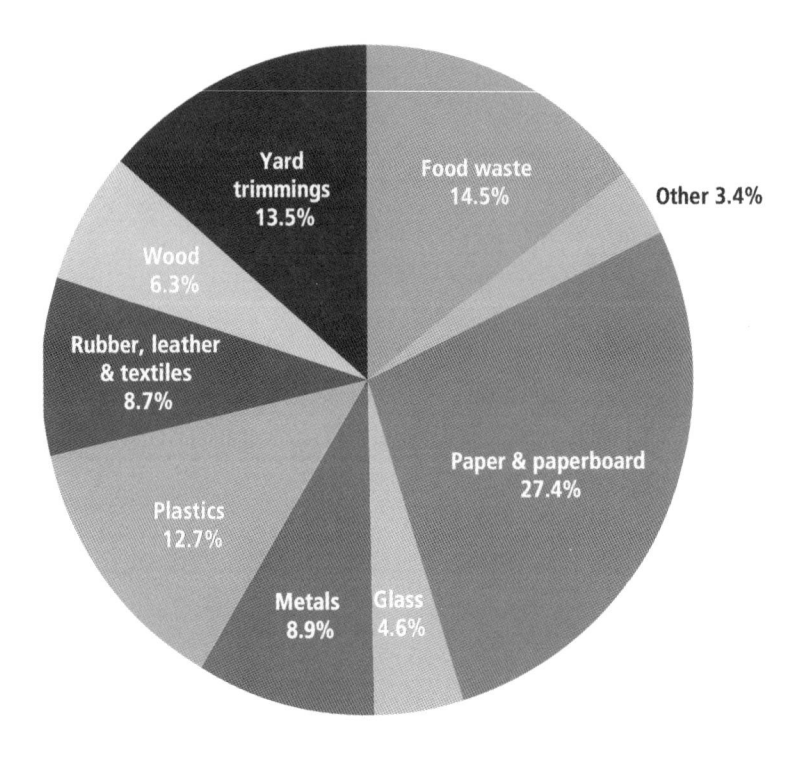

Figure 7. Total municipal solid waste generation by material in 2012 before recycling (US Environmental Protection Agency 2014b).

more powerful than carbon dioxide (Gunders 2012). Food scraps have high moisture content, so they decay faster than other organic and inorganic materials, and as a result, they produce a disproportionately large amount of methane. Decomposing food waste represents 90 percent of a landfill's methane emissions, and landfills accounted for 18.2 percent of all greenhouse gas emissions in the United States in 2012 (Gunders 2012; US Environmental Protection Agency 2014a). It is important to note that this is a natural process that also occurs during composting, but turning or aerating the compost pile replenishes it with oxygen, which mitigates the amount of methane produced (Cooperband 2002). As a result, the United States needs to manage its waste in a more sustainable manner.

Composting operations vary in size and technique, and these factors influence how governments integrate composting into urban areas. Table 6 provides

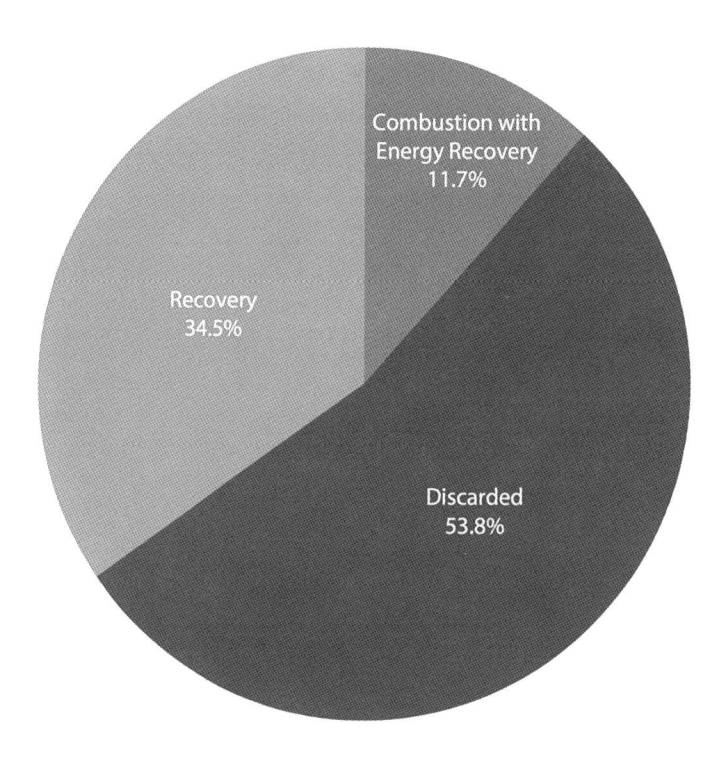

Figure 8. Management of municipal solid waste in the United States in 2012 (US Environmental Protection Agency 2014b).

a list of common composting terms. Aerobic in-vessel and anaerobic digestion facilities can accommodate large amounts of organic waste and require a substantial amount of land, while pile and holding-unit composting are more appropriate for small amounts of organic waste because they use manual labor to aerate and manage the material. Categorizing the types of composting into large and small-scale operations creates two basic approaches to promoting its use in urban areas: centralized and decentralized. Centralized composting handles a large volume of organic material at a single facility and can be part of an organized collection program. In contrast, decentralized composting refers to individual activities that smaller operations manage for their own use. As a result, the composting processes serve different functions within a community. Like most, if not all, urban problems, organic waste disposal is

Table 6. Common composting terms

Term	Definition
Composting	Controlled decomposition, which is the natural breakdown process of organic material.
Aerobic in-vessel	An enclosed composting system that allows the operator to control the temperature, oxygen, and moisture. A common system for commercial compost producers because it permits the greatest amount of environmental control.
Anaerobic digestion	A process that breaks down organic materials without oxygen. The process occurs in large- and small-scale composting systems, but the method produces biogas, which is a type of fuel, and compost so it is an efficient method for commercial or industrial operations.
Pile	A composting method that does not require a structure. The operator simply places organic materials in a mound, i.e., a pile, and turns the pile as necessary. The operator does not have to turn the pile to create compost.
Holding unit	A structure that either partially or completely encloses the organic material that the operator turns on a regular basis. The ideal size is 3x3 or 5x5 feet. Holding units are commonly known as bins.

Source: Cooperband (2002).

complex and one intervention will not be a panacea, which makes it necessary to foster both systems (City of Austin 2011; Buzby, Wells, and Hyman 2014).

Local governments play a vital role in promoting both centralized and decentralized composting because their primary function is to promote the health, safety, and welfare of the community. One of the tools they use to accomplish this is their municipal code. This code controls every function of a city, which includes the waste management system and land-use regulations. Cities can encourage or discourage specific activities through the ordinances in their municipal code, and therefore, by adopting policies that promote centralized and decentralized composting, they will have the most impact on diverting organic materials from landfills (City of Austin 2011).

Limitations

The regulatory methods may apply to rural communities, but cities are the primary audience for these recommendations. Composting for agricultural uses in rural areas is a separate issue that is outside the scope of this book.

STRUCTURE OF COMPOSTING REGULATIONS

The interplay between federal, state, and local municipal solid waste laws can be complicated, and historically, governments have defined and treated organic and inorganic waste according to the same standards, that is, as a commercial disposal service (US Environmental Protection Agency 2013a, 2014c). However, it is necessary to distinguish between the historical intent of municipal solid waste laws and the ability of composting to safely support sustainability initiatives such as zero waste and urban agriculture (Ohio Environmental Protection Agency 2012; City of San Francisco 2009).

The federal government has regulated waste management since 1965, but Congress has amended the regulation several times in order to respond to the growing prevalence of waste in our society. Congress significantly expanded the scope of the Resource Conservation and Recovery Act (RCRA), which was enacted in 1976, by establishing programs for different types of waste, such as solid waste and hazardous waste, as well as for underground storage tanks, and by creating management and disposal standards for each program (US Environmental Protection Agency 2013c). The standards do not preempt the states' power to regulate solid waste. Composting is a method to dispose of organic solid waste, so it is subject to state authority.

RCRA, however, influences the extent to which states use composting as a sustainable disposal method because it defines food scraps as a component of municipal solid waste and most governments manage the material in a similar manner. According to RCRA, municipal solid waste includes "everyday items we use and then throw away, such as product packaging, grass clippings, furniture, clothing, bottles, food scraps, newspapers, appliances, paint, and batteries" (US Environmental Protection Agency 2014b). The regulatory framework for solid waste varies significantly from state to state, which means that each state has different requirements for recycling and composting specific materials. But, most, if not all, regulate the activity by controlling the siting, permitting, and operations of the disposal facilities.

Unlike traditional and recycled waste streams, state composting regulations have two components: (1) the siting, permitting, and operation requirements for regulated activities, and (2) a list of exempt activities. The standards and requirements for the regulated and exempt operations vary according to each state. Most states, however, limit the exempt activities by restricting (1) the type of materials to yard trimmings and food scraps, (2) the source of materials and use of the product to the site in which it was generated, and (3) the size of the pile or facility. Again, the size limitation varies by state, but at a minimum, most states exempt household composting and some specify an amount, for example, 500 tons per year (Purman 2008). For regulators, the objective of these thresholds is to exempt operations that landowners can maintain in a nuisance-free manner, which typically include smaller operations. This chapter associates the regulated operations with centralized facilities and the exempt activities with decentralized methods.

Several states have amended their composting regulations to decrease the procedural requirements for regulated operations and to increase the size thresholds for permit-exempt activities. These regulations also expressly list food scraps as a permissible material. There are two approaches to increasing the thresholds for permit-exempt activities: (1) exempt operations according to the total land area dedicated to the use on a parcel, and (2) exempt activities up to a specific size, for example, 50 cubic yards or 10 tons per week (State of Ohio 2012; State of Wisconsin 2012; State of Massachusetts, n.d.). Illinois also amended its Environmental Protection Act to establish that an operation composting "food scrap, livestock waste, crop residue, uncontaminated wood waste or paper waste" is not a pollution control facility; that is, it is not subject to the same standards as everyday waste facilities (State of Illinois 2013). These regulations demonstrate that states are recognizing composting as a sustainable disposal method that works with the traditional municipal solid waste stream, and encouraging local governments to figure out an alternative solution to managing organic waste (Arroyo-Rodriguez and Germain 2012).

Local governments can address this directive by integrating centralized and/or decentralized composting operations into their community. Since the methods are based on the scale of the activity, the city may address each initiative through different departments and areas of the municipal code. The waste management or public works department is usually responsible for collection regulations and operations, so a city may administer its centralized composting program through the same area of the municipal code, and the planning department addresses land-use regulations through its zoning code, which corresponds to the individual nature of decentralized composting activities. Therefore, each department has a unique relationship to the state regulations and will have specific considerations. Regardless of their traditional duties and specific roles, the departments should work together to develop cohesive composting regulations and programs because it will help the city establish a mutually supportive approach to urban composting (Evans-Cowley and Arroyo-Rodriguez 2013).

Because of the large scale of centralized composting operations, the facilities are subject to state siting, permitting, and operation requirements. Local governments either directly or indirectly interact with these regulations when they establish curbside organic collection programs because they require separate processing facilities. As a result, the municipal solid waste collection agency—for example, the city or an independent contractor—will either build a new facility or use an existing composting operation. San Francisco was the first city to establish a curbside organic collection program, and since it started in 2009, several local governments have been exploring similar initiatives. More than five cities have implemented a collection program, and they have experienced a positive impact on their landfill and support from their community. Therefore, it is reasonable to believe that more cities will integrate curbside organic collection into their waste management system.

Regarding decentralized composting, local governments can promote the activity by establishing it as a permitted use throughout their jurisdiction. This is a significant change to most zoning codes because cities typically limit composting to residential properties and impose strict size and setback requirements. However, it is important to divert organic waste from landfills because most of the zoning codes that explicitly permit composting outside residential districts associate its use with community gardens and urban farms, but most of the urban-agriculture zoning regulations do not address composting. Given the natural connection between the two uses and the growing prevalence of urban-agriculture regulations, decentralized composting is an excellent method of integrating the activity throughout a city and supporting sustainable disposal initiatives (Ohio Environmental Protection Agency 2012; Arroyo-Rodriguez and Germain 2012). The following sections explain important factors to consider when implementing centralized and decentralized composting regulations.

CENTRALIZED COMPOSTING

Centralized composting is a sustainable waste management method that local governments can use to divert most of the city's organic material from the landfill because the capacity of the facilities and the disposal and collection practices mimic those of the traditional waste management system. One of the major benefits of large-scale composting facilities is their ability to process large amounts and varieties of organic materials. The high temperatures of in-vessel and anaerobic facilities can break down organic materials that are unsafe or too dense for small composting systems and that consequently end up in the traditional waste stream. The second major benefit to centralized composting is that the disposal and collection methods for the residents and waste haulers are very similar to current practices. These characteristics make centralized composting a viable alternative to landfills.

The city of Madison, Wisconsin, recently explored the feasibility of implementing separate organic and inorganic collection programs. According to its study, approximately 75 to 85 percent of the community's waste is organic material that is suitable for recovery in a large-scale composting facility. It found that one anaerobic digestion facility could accommodate all of the source-separated organic waste currently collected by the city and could accept waste from private haulers and surrounding communities in order to achieve an optimal economy of scale (Organic Waste Systems 2012). Source-separated organic waste refers to compostable materials that are separated from other waste streams when they are disposed of, and this occurs at the source, as opposed to at the waste management facility. Based on its ability to process additional source-separated organic waste in the immediate and long-term future, the city decided to pursue an organic collection program. At the time of the study, Madison's population was 233,000, so it demonstrates that one centralized facility has the potential to recover a significant amount of organic waste.

Several cities have already recognized this opportunity and have adopted mandatory composting regulations to engage people within the jurisdiction in their diversion efforts. Mandatory composting is commonly known as "municipal composting," but some cities contract their waste collection services with a private company so it is not always an appropriate term. This chapter replaces "municipal composting" with "curbside organics collection" because the latter term encompasses both municipal and private waste collection agencies. All curbside organics collection programs have two fundamental requirements: (1) residents must separate compostable, recyclable, and landfill materials, and (2) the waste collection entity must establish a means to collect and process the organic material. Beyond these characteristics, the implementation methods will vary according to the city and its waste management system.

This section identifies the factors that local governments should consider when integrating curbside organics collection throughout their community and into their municipal code. Successful initiatives have used a process-oriented approach that changed the way residents view and dispose of their waste. Even though the day-to-day management of source-separated organic waste is similar to that of current disposal practices, it involves a government-mandated lifestyle change, and people can approach such changes with resistance. As a result, it is necessary to base mandatory composting in the community's public and environmental welfare. Here, this involves recognizing waste reduction as a city concern in formal resolutions, articulating the resolution as a solution in city plans, and developing regulations to implement the solution.

Zero-waste resolutions

All cities that offer curbside organics collection adopted a zero-waste resolution before they established their program. They used the resolution as the rationale to develop waste management studies and integrate sustainable practices into their plans and regulations. San Francisco, California, was one of the first cities to adopt a zero-waste resolution. In 2002, the resolution was innovative in and of itself, and consequently, it has become a model for similar initiatives. Every resolution is unique to its own community, but they all have four common goals:

1. Connect the initiative to sustainability goals and/or principles.
2. Identify the problem, which is usually the impact of landfills on the environment.
3. Set a goal to divert a percentage of landfill material by a specific date.
4. Identify a department to implement the resolution or solicit a study to analyze the appropriate solutions.

Cities typically differ in how they address the third characteristic. For example, San Francisco took an action-oriented incremental approach, and Austin, Texas, adopted a planning-based comprehensive approach to achieving zero waste. San Francisco's resolution set a goal to divert 75 percent of its waste from the landfill by 2010 and a requirement to establish a zero-waste date once it reaches 50 percent (City of San Francisco 2002). The city took specific actions to accomplish its goal, including passing the Mandatory Recycling and Composting Ordinance in 2009.

The city of Austin passed its zero-waste resolution in 2005, which required the city manager to conduct a study of the waste economy and develop a solid waste services master plan (City of Austin 2009). The objectives of the study and the plan were to help the city identify and pursue effective steps to achieve its goal of zero waste by 2040. This allowed it to evaluate all aspects of its waste stream and implement changes throughout the system. The city passed a *Zero Waste Strategic Plan* in 2008 and a *Resource Recovery Master Plan* in 2011 (City of Austin 2008, 2011). According to the master plan, curbside organics collection is one of several methods being used to integrate sustainable waste management throughout the community.

At this point, it is not clear whether one approach had better results, but it is reasonable to believe that a planning-based approach will help the city inform residents about the importance of sustainable waste management and garner their support. Additional best practices will emerge as more cities adopt zero-waste resolutions.

City plans

Most cities develop a waste management plan to ensure that they have the capacity to dispose of all of the waste generated in their community. The analysis is comprehensive because it addresses waste generation, disposal and recovery, collection, and management, but the recommendations have historically focused on landfills as the primary waste management method (Evans-Cowley and Arroyo-Rodriguez 2013). However, pursuant to the zero-waste resolutions, several cities have integrated sustainability principles into their formal analyses and management strategies. Composting is a common sustainability recommendation because it is a method to recover organic waste and reuse the material as a soil amendment. Therefore, based on the increasing trend toward zero waste and the benefits mentioned above, it is necessary for waste management plans to shift their attention from landfills to composting.

Austin adopted a multifaceted waste management strategy in its *Resource Recovery Master Plan*. The plan is a framework for how the Resource Recovery Department provides services to the community, and it redirects the city's perspective from managing waste streams to managing individual materials, meaning that trash is viewed as a resource with a second life. Austin applies this principle to managing organic materials by recommending a variety of collection and processing methods. Regarding composting, the plan establishes a list of highest and best-use practices that organizes the main methods according to their ideal use—for example, individual holding units are best for managing food scraps and anaerobic digestion is an end-use disposal option with the ability to capture energy (City of Austin 2011). This allows the city to integrate centralized and decentralized composting throughout the community.

Given the scope of establishing and implementing centralized composting services, the plan specifically evaluates the components of a citywide organics collection program. It identifies (1) the existing organics collection services, (2) the role of the city and private service providers and haulers, (3) the process for implementing a program, and (4) the main components of a composting ordinance. This information not only provides the preliminary analysis for starting a curbside organics collection program but also develops the rationale for initiating the service.

Mandatory composting regulations

The final stage of establishing a curbside organics collection program includes adopting a mandatory composting ordinance because it provides the legal basis to implement the program. As mentioned above, cities manage their municipal solid waste collection services in different ways, so the means to implement a curbside collection program will vary, but an effective regulation should do the following:

1. Require people to separate organic materials and participate in a curbside collection service.
2. Articulate collection standards and expectations.
3. Establish enforcement mechanisms and penalties.
4. Set a program effective date.

A city may need to amend several different sections of its municipal code and possibly adopt additional administrative rules in integrating the criteria. For example, San Francisco's Mandatory Recycling and Composting Ordinance amended its Environment Code, Public Works Code, and Health Code (City of San Francisco 2009). Portland regulates the collection, processing, and end use of municipal solid waste through its Solid Waste and Recycling Collection Chapter but articulates the standards for collection agencies as an administrative rule (City of Portland 2009).

DECENTRALIZED COMPOSTING

Since its resurgence in the late twentieth century, the local food movement has focused primarily on establishing and promoting methods that facilitate the production of food closer to consumers, but it has largely ignored the intrinsic connection between urban agriculture and sustainable disposal practices. Rural farmers compost vegetative waste and animal feces to recycle the material and restore nutrients to their soils. When cities integrate composting into their jurisdiction through urban-agriculture regulations, they can facilitate the same results but at a community level.

When a city explicitly addresses composting in its zoning code, the primary purpose is to develop regulations that protect public health and the environment and promote effective composting (Purman 2008). Cities regulate all land uses according to specific criteria, and the following subsections identify the zoning requirements that are vital to urban composting, including consistency with state requirements, use classification, definition, permissible and compost application, site restrictions, and registration. Each subsection will identify the important aspects of the criterion and provide examples of how other municipalities have addressed it. The criteria and content are based on a review of zoning codes in states that recently updated their regulations to promote composting. I conducted the review between August 2013 and June 2014 with the objective of identifying cities that address composting as a nonresidential land use and understanding how they regulate the activity.

There are different methods of integrating any activity into a municipal code, so every city should tailor its approach to meet its regulatory framework and community interests. Before a city develops criteria for composting activities, it needs to identify whether its code recognizes composting, and if so, how it regulates and where it permits the activity. This will help the city staff understand what they need to do to effectively and efficiently integrate the activity into their code.

Consistency

Local composting regulations should align with state requirements. This will not only avoid confusion among government regulators, enforcement agencies, and residents but will also ensure that local uses have the full benefit of state exemptions. For example, Ohio exempts composting activities with an aggregate area of 300 square feet, but the Cincinnati zoning code says that "a maximum area of 200 square feet may be used for composting." Furthermore, Ohio allows landowners to compost materials that they did not generate on their site, but Cincinnati expressly prohibits this practice (State of Ohio 2012; City of Cincinnati 2010).

The differences between these state and local regulations are minor, but inconsistencies can impact a landowner's perception of the regulatory burdens associated with the activity, which will ultimately influence whether he or she composts. Local governments can avoid this misconception and promote transparency by synchronizing the state and local thresholds for urban composting. This practice would not inhibit their fundamental ability to go beyond the state regulations.

Use classification

There are two primary methods of permitting composting in zoning codes: (1) establish composting as an accessory use in specific zoning districts, and (2) describe composting as part of a primary use category—for example, community gardens or another use that benefits from on-site composting. Accessory uses are incidental to the primary use and may have additional size and setback standards, and primary uses are allowed by right and must meet the specifications of the zone. In either approach, the city should list compost piles and bins as acceptable accessory structures. Madison regulates composting according to the first method. More specifically, it permits composting as an accessory use in several zoning districts throughout the city and has separate permissions for various agricultural uses (City of Madison 2013). The city permits composting wherever it allows urban-agriculture activities, and in doing so, it recognizes that composting is an appropriate accessory use to community gardens.

The second method permits composting by imposing additional developmental regulations on the primary use category and outlining the composting standards in that section. For example, Cincinnati's zoning code permits community gardens as primary or accessory uses in several districts throughout the city, subject to additional standards. These use-specific standards clarify that composting is included as part of a community garden use and outline limitations on composting activities (City of Cincinnati 2010; Ohio Environmental Protection Agency 2012). In effect, the second approach directly recognizes the inherent relationship between composting and urban agriculture and ensures that the activities support each other.

Several cities allow community gardens and urban farms throughout their jurisdiction but do not address composting in their zoning code. In this case, a city can integrate the use by establishing it as an accessory use and permitting it in the districts where it allows urban-agriculture uses. However, the municipal code may address composting in its municipal solid waste and health and safety sections, so the city should address every aspect of the code that relates to the proposed changes and apply consistent principles and language.

Definition

Every zoning code includes a definitions section to clarify terms and reinforce the intent of the regulations. If a local government is looking to encourage composting, it should define *composting* as a use or activity and *compost* as the product of the use or activity. Composting definitions vary among government agencies and departments, but in order to promote the activity as a sustainable disposal method, it should include three factors:

1. The definition should convey the management criteria—for example, turning the piles to avoid odors and vermin nuisances (Ohio Environmental Protection Agency 2012).
2. The city should either tailor the definition to the intensity and type of composting that it wants to encourage, or address large-, medium-, and small-scale activities.
3. The definition should include food scraps as a component of the activity and product to ensure that it promotes composting as a method to mitigate landfill and food waste problems.

For example, the EPA and the Boston Zoning Code provide definitions of composting that include the three criteria. The EPA states that composting is the process of "combining organic wastes (e.g., yard trimmings, food scraps, manures) in proper ratios into piles, rows, or vessels; adding moisture and bulking agents (e.g., wood chips) as necessary to accelerate the breakdown of organic materials; and allowing the finished material to fully stabilize and mature through a curing process." Compost is defined as "organic material that can be used as a soil amendment or as a medium to grow plants" (US Environmental Protection Agency 2014d). The city of Boston combines its definition for composting and compost into one term and integrates the criteria in a more concise manner. There, composting "is a process of accelerated biodegradation and stabilization of organic material under controlled conditions yielding a product [that] can safely be used as a fertilizer" (City of Boston 2013). Local governments can modify or combine existing definitions, such as those of the EPA and city of Boston, to ensure that their regulations contain the three factors and meet their needs.

Permissible materials and compost application

The zoning code should contain a comprehensive list all of the materials that landowners can use in their composting operation, the source of the materials, and the application of the final product. These criteria will dictate the role and effect of composting in a city because they connect the activity to other functions within the municipality, such as urban-agriculture uses and waste disposal operations. Despite their best efforts, it is likely that most of the existing compost regulations will have a minor impact on these functions because they limit the permissible materials to organic waste generated on the site of the composting activity. This severely restricts the variety of compostable materials and quantity of compost that an operation can produce, and as a result, the benefits will be negligible (Arroyo-Rodriguez and Germain 2012). Cities can promote composting as an appropriate urban land use and a sustainable disposal method if they specifically list the permissible compost materials, their source, and the application of the product in their zoning code.

Given the sensitive nature of composting, cities should consider several factors when drafting this portion of the ordinance. Important components include the following:

1. Synchronizing the thresholds with state requirements.
2. Recognizing that all organic materials may not be appropriate in all districts—for example, animal manure may cause a noticeable odor in a downtown or residential district (Ohio Environmental Protection Agency 2012).
3. Limiting composting materials to source-separated organics (Harrison and Richard 1992).
4. Accepting composting materials from off-site sources (State of Ohio 2012).
5. Permitting the use of the product on off-site locations (State of Ohio 2012).

This section should be very specific because vague criteria could lead to unintended interpretations and potential nuisances.

Site restrictions

Composting operations are subject to the size and setback requirements of either the zoning districts' accessory-use standards or the development criteria specific to the primary use. As previously mentioned, the composting area size limitations should be consistent with state regulations. The setback requirements differ from the size limitations because they influence the location of the compost pile or bin on the property. Composting is a sensitive land use, so cities should tailor their setback requirements to their needs. Further analysis of why cities specify the distances they do will help us understand whether setback requirements are related to other variables such as population density.

Setback requirements for decentralized composting activities vary by community. Some cities establish setbacks specifically for composting activities, and others subject composting to district-wide setbacks for all accessory or primary uses or structures. When composting is permitted in association with an urban-agriculture use, it may be subject to additional locational restrictions. For example, both Cincinnati and Dayton have established setback requirements specifically for composting activities (City of Cincinnati 2010; City of Dayton 2010). Boston and Madison permit composting as part of an urban-agriculture primary use and apply the setback requirements of the base zoning district (City of Boston 2013; City of Madison 2013). Chicago, on the other hand, does not specify setbacks for urban-agriculture accessory uses and limits composting in association with urban-agriculture activities through size limitations that apply to small-scale composting activities (City of Chicago 2014).

Registration

If a city permits composting as an accessory permitted use, it does not need to develop a registration system, but the information would help the city monitor the use of composting and provide valuable data on its effects throughout a community. Registration information would enable the city to observe composting operations to ensure that they comply with basic health and safety standards. Such data would also be vital in analyzing the impacts of composting operations and trends in municipal solid waste because states do not require registration or notification for exempt operations. This practice has made it difficult to quantify the amount of organic waste diverted from landfills to composting (Plat, Ross, and Poland 2012). None of the cities in the survey sample have a registration program for decentralized composting uses, but if a community is particularly concerned with tracking its waste diversion, it may be worth considering.

The city of Boston has a specific review procedure, Comprehensive Farm Review, for more intensive urban-agriculture uses, and since the city permits composting as a component of the primary permitted use, the review analyzes the composting operations. Comprehensive Farm Review allows the city to have more oversight on the design of the proposed operation and increases its awareness of the number and extent of agricultural activities (City of Boston 2013). However, it is important to identify that this is a review procedure and not a registration system, and as a result it has two limitations. First, if the city wants to maintain a record of the Comprehensive Farm Review composting operations, it needs to transfer the review application information into a database that can track the total number of composting activities within its jurisdiction. Second, it would gather information on composting activities associated only with more intensive urban-agriculture uses.

If a city pursues compost registration, it should consider the capacity of landowners and its staff to effectively implement and manage the program. The registration process should be simple and inexpensive for landowners, or in other words, it should not discourage them from composting organic waste. The database should also be easy to maintain in order to minimize monitoring by city staff.

CONCLUSION

The solid waste disposal statistics demonstrate that local governments control a large portion of the organic materials generated within their communities through traditional municipal solid waste practices. Based on landfill emission statistics, this approach is problematic. In order to transform this problem into an opportunity, municipalities must integrate laws of different types with organizational activities and opportunities.

Reconstructing the current situation will enhance the multifunctionality of cities and regions in three ways. First, it will enhance their resilience by mitigating environmental problems. Second, it will provide new economic opportunities that support urban agriculture. And third, further research could substantiate additional benefits of restructuring municipal solid waste systems, such as energy production from biodigestion. Therefore, by decreasing the procedural requirements for composting facilities and increasing the thresholds for permit-exempt operations, several states have recognized composting as a sustainable disposal method and the need for local initiatives that divert food waste from the landfill. Cities can and should take advantage of this deregulation by integrating centralized and decentralized composting processes into their waste management systems and land-use regulation.

DISCUSSION QUESTIONS

1. Although some states are loosening their restrictions, all states do have significant restrictions controlling waste processing. Is merely decreasing procedural requirements the answer for states, or are there further actions they can take to enable composting by urban farmers?
2. Do cities and city departments have the capacity to host optimal composting while keeping compost out of landfills, as in Madison, Wisconsin? What does it take for a city to do this successfully?
3. What role should ordinances like San Francisco's and Austin's zero-waste resolutions play in inspiring other cities to pursue similar legislation? What gives these resolutions "teeth"?
4. The city of Cincinnati has been successful in promoting composting, but it has legislation that conflicts with that of the state of Ohio. On whom does it fall to avoid conflicting legislation?
5. Suerth states that city planners ought to set very specific definitions for compost and composting. Does this requirement for specificity rely solely on planners, or with other departments and actors?
6. Suerth's chapter discusses primarily the role of governments and government actors in compost and composting. However, she makes scant mention of urban farmers. Do urban farmers have a role to play in optimizing access to and production of compost?
7. Are there other barriers to urban composting outside of government that Suerth may not have touched on? Social barriers? If so, how do these barriers play out, and what can be done about them?

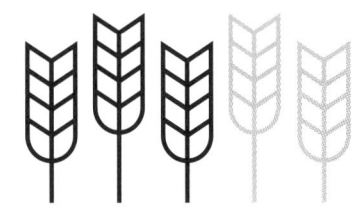

SECTION THREE
Production

Within urban agriculture, a discussion of production encompasses those factors that directly influence growing food in the city. In chapter 7, Silva and Pfeiffer examine the interactions of the biophysical, technical, and socioeconomic components of the farming system and explain how the urban agroecological system shapes and is shaped by the soil conditions, infrastructure resources, workforce, and business structure of urban farms.

Production innovations can also include nontraditional growing locations, including planning for temporary use of land and taking advantage of unique aspects of the urban environment and the involvement of communities in producing food. In chapter 8, Companion documents the empowerment of urban-dwelling Native Americans through container gardening and emphasizes how food and herb production fosters cultural continuity and individual health by connecting gardeners to traditional cultural and religious practices. Remembering these potential connections and concomitant relationships is central to realizing the potential of urban food production and food systems.

In chapter 9, Lawson, Drake, and Fitzgerald showcase a farmers market initiative in an area of New Brunswick, NJ with low food access and show that developing such markets requires a long-term commitment by both the community and institutional or governmental partners, as they may not be as financially viable as markets in more affluent areas.

Leadership by members of food-insecure communities is more visible in urban agriculture today than historically, and we see more examples of shared leadership between those in positions of privilege seeking to make the food system more just and those from communities facing injustice in their food systems. However, the risk of paternalism and tensions among communities still exists, as chapter 10 illustrates, and this risk requires vigilance among activists and scholars alike.

CHAPTER 7

Agroecology of Urban Farming

ERIN SILVA AND ANNE PFEIFFER

Urban farming and the associated marketing and food distribution framework create unique agroecosystems. Agroecosystems can be defined as ecological systems modified by human beings to produce food, fiber, or other agricultural products. The study of agroecology provides one lens through which to evaluate urban-agriculture production—or, as per the definition above, the urban agroecosystem. Agroecology uses ecological theory to study, design, manage, and evaluate agricultural systems, considering interactions of biophysical, technical, and socioeconomic components of farming systems (Altieri 2014).

The focus of this chapter will be to discuss the current status of urban-agriculture production in the United States using an agroecological framework. In order to create a boundary as to the type of farm included in this analysis, the evaluation will be limited to operations that are located within or closely proximate to a metropolitan area and produce food on a commercial or community scale with the intention that the products be consumed in the same geographic area. Urban growers overcome many challenges with innovative solutions specific to their location and the goals of their particular project. Our examination of urban agriculture through the lens of the agroecological model presents each agroecological principle as it relates to urban agriculture, followed by a discussion of the relevant challenges and resulting innovations.

PRINCIPLES

Gordon Conway outlined a framework with which to evaluate the performance of agroecosystems in his paper "Agroecosystem Analysis" (Conway 1985). He proposed that agroecosystems be described by four properties—productivity, stability, sustainability, and equitability. These properties describe the dynamic functions of an agroecosystem and its associated agronomic, social, and economic components. Factors and processes affecting the specific nature of these properties in a given agroecosystem—for example, crops produced, availability of resources, skill of labor pools—are described in the context of their individual and interrelated characteristics. The factors and processes may have overlapping influences on the primary system properties, exhibiting both synergistic and antagonistic relationships to such properties.

Depending on the nature of the influences on a particular agroecosystem, the precise metrics used to characterize each of these properties will vary. For many agroecosystems, productivity is measured through the quantification of the food, fuel, or fiber produced for human use by farming practices. Productivity can thus be influenced by fertility management approaches, the selection of crop varieties, the quantity and quality of the land base, and other agronomic factors. The sustainability of an agroecosystem evaluates the ability of that system to maintain a specified level of production throughout a longer-term outlook; thus, an evaluation of sustainability includes an analysis of the forces that could cause major production disturbances. Conversely, an evaluation of the stability of an agroecosystem takes into account the ability of that system to consistently produce a product in the face of small disturbing forces arising from the physical, biological, social, and economic environment. The equitability of an agroecosystem is the ability of the system to share agricultural production fairly among the impacted population; it evaluates the evenness of product distribution throughout the agroecosystem among the human beneficiaries. Through this framework, one can identify the important factors and processes that affect the primary system properties.

Productivity

The productivity of an agroecosystem can be described by several metrics. Most commonly, the productivity of an agricultural system has been measured by yield as the amount of food that a system can produce per unit of land. Alternatively, productivity could be determined by other calculations, such as the potential yield of a system per unit of energy or per unit of labor. Although not a standard metric in most traditional agricultural analyses, productivity could also include measurements of secondary benefits, such as job training, skills building, or the creation of other external benefits. The determination of the most appropriate metrics to use in a given system is particularly important in the urban-agriculture context, where many urban farms strive to accomplish (often with success) alternative community and social goals beyond the production of food. The evaluation of a system's productivity based on social factors is an example of the interrelatedness of agroecosystem principles and argues for the need for a systems-based evaluation of urban agriculture.

In an agroecological context, the characteristics of the agronomic system can significantly influence potential yields. These characteristics include soil fertility and quality, availability and skill of labor, and seasonal disease and pest pressure. Many of these factors that influence productivity also influence system properties, highlighting the interrelated nature of factors included in this framework.

Soil quality and fertility Soil quality is a critical factor that influences the productivity of any agricultural site. Many urban farms are sited on vacant lots where soil has been significantly altered by human activity such as previous building construction, parking lots, or waste sites. As a result, soil quality, as determined by fertility, microbial activity, structure, and drainage, is frequently suboptimal and in need of remediation. The presence of soil contaminants resulting from previous activities and their potential impact on food safety are also relevant when describing the condition of soil quality on urban farms.

As a component of soil quality, optimal soil fertility is a particularly important factor in achieving the maximum potential productivity from any agricultural operation. In urban environments not easily served by farm cooperatives or other agricultural supply stores, urban farmers may have difficulty sourcing readily available inputs at an affordable cost. As such, fertility inputs may need to be imported into the urban environment from the surrounding rural community. Because of their weight and volume, costs to ship these items can be prohibitive. In addition, urban farms may not have access to freight carriers or infrastructure, such as forklifts and loading docks, necessary for the shipment and delivery of heavy or bulky agricultural inputs.

Urban farmers have successfully adopted a variety of strategies to improve soil quality and thus increase the productivity of their farming operation. Given the challenges of poor soil and difficulty in importing fertility amendments, urban farms have come to rely heavily on compost to improve soil quality, supply crop nutrients, and build soil structure. Composting efforts in an urban agroecosystem are supported by the close proximity of restaurants, food-service facilities, grocery stores, and food processors that can supply abundant pre- and post-consumer food waste. Additional sources of compost feedstock are available through landscaping activities and from waste generators and haulers, who are often amenable to dumping their waste at urban-agriculture sites rather than paying fees at conventional disposal sites. Urban farms have also adopted the process of vermicomposting (producing compost using red wiggler worms) to efficiently produce a rich soil amendment with high levels of plant-available nitrogen (Atiyeh et al. 2000).

Though compost plays a central role in improving soil quality and fertility on urban farms, challenges with respect to compost production and quality are common. Depending on initial feedstock and production techniques, compost quality can be highly variable and, in some cases, may not be adequate to supply crop nutrient needs or may introduce contaminants into the agroecosystem. For example, readily available feedstocks such as yard waste may not be desirable for use as compost because of concerns about herbicide carryover and weed seeds in the final product (Brown and Jameton 2000). Despite the availability of resources to produce compost in urban environments, not all urban growers have access to sufficient quantities. With few independent compost vendors, urban farms without the necessary time, labor capacity, or space to produce compost on-site may have difficulty obtaining the product. Some municipalities or land-use agreements prohibit on-site composting or restrict composting to materials collected on-site, further limiting growers from generating enough compost to meet their production needs.

Urban farmers have integrated soil improvement using compost into their farming practices through varying strategies. Existing urban soils present on a farm site are often amended by using compost or importing new soil. These continued additions of compost can, over time, improve both the fertility and quality of degraded soils by increasing soil organic matter, improving soil structure and tilth, and restoring soil microbial activity.

As an alternative to using compost to amend large areas of urban farm sites, some operations have limited the use of imported soil and compost by growing plants in pots or small raised beds, thus avoiding larger-scale soil amendment. While requiring less capital investment than removing pavement and remediating or importing soil, pots and raised beds also provide less growing space, thus limiting the potential yield per unit area. Labor requirements to produce and harvest crops in these systems may be high, as pots often need to be moved on a regular basis and must be individually tended by hand. Because of less insulation from soil temperature fluctuations, plants grown in pots outdoors or in raised beds atop asphalt will be more susceptible to heat and cold stress than in-ground planting, limiting plant growth. In addition, nutrient and water management may be more challenging than in an in-ground planting system, decreasing the farmer's ability to achieve the yield potential of a crop.

Selection of crops and livestock   Crop selection greatly influences the overall productivity of urban farming ventures. The space-limited environment of many urban farms drives creative use of space, influencing the array of crops grown. Root crops and vining crops, which typically require more vertical and horizontal space for production, are often not grown except in an educational or demonstration garden or when integrated into a specific marketing strategy. Crops well suited for dense spacing, intercropping, and container planting (e.g., leafy greens and tomatoes) are more common among space-intensive urban growers. Some urban farms, such as Growing Power in Milwaukee, Wisconsin, take dense seeding a step further by overseeding—as lettuce and other greens mature, more seeds of the same crop are broadcast on top of the existing crop to create continuous production. To be successful, these practices require the maintenance of high soil fertility and careful visual scouting to monitor insect and disease pressure. With greater yield potential per square foot, these techniques tend to be more labor intensive than traditional crop spacing, representing a trade-off between maximizing productivity and limiting labor inputs.

Livestock production on urban farms occurs less frequently than the production of crop plants. However, some urban farms have taken a creative approach to livestock production with the integration of aquaponics. The use of aquaponics creates integrated systems in which fish are cultivated in managed tanks and the resulting fish manure fertilizes plants grown using water from the tanks. Such a system enhances productivity per unit of land by producing a high-quality protein source in addition to the vegetables. As this system most commonly employs hydroponic techniques in indoor environments, the vegetables grown using these systems (often baby leafy greens and herbs) are exceptionally clean and typically of very high quality, optimizing their sales in high-value markets. Despite potential positive impacts on productivity, the financial viability of aquaponics systems remains uncertain because of the relative newness of the technology and lack of documented business models.

Challenges of integrating social goals Urban-agriculture projects are characterized by diverse missions that often embed food production into one or more social goals, such as community empowerment, nutrition, youth engagement, education and job training, or neighborhood beautification (Mogk, Wiatkowski, and Weindorf 2010; Hagey, Rice, and Flournoy 2012). Though food production remains a central focus for many operations, it is often a means to achieve other social benefits rather than the singular goal. The integration of food production with social goals may result in reduced agricultural productivity, but overall productivity may be high when social outcomes are considered in an integrated analysis. An evaluation of the productivity of urban agriculture must therefore account for not only the food output but also the social goods produced by a given project.

Social missions can compete with production goals and crop yield because of a lack of skilled labor, a need for increased supervision, modified crop selection, and increased training requirements. Skilled farm labor is often scarce on urban farms because of both financial and logistical reasons. When organizations with broad missions have a limited number of staff, many of whom hold diverse organizational responsibilities, it is often difficult to recruit employees with the ideal agricultural production experience. Farm managers on urban farms with social missions often have experience working in nonprofit organizations but lack farming knowledge. This issue can be exacerbated if hiring and management decisions are made directly by a Board of Directors with no direct agricultural experience.

Urban farms with social missions often prioritize hiring workers or encouraging volunteers from the community. Along with increasing food availability and enhancing food security, these practices enhance employment and skill building in local communities, an additional benefit of urban farms. However, with respect to maximizing production potential on a farm, these workers often lack practical, hands-on farming experience, hindering the ability of urban farms to reach their yield maxima. Many farms with a high proportion of unskilled labor find it necessary to select crops based on the ability of their labor pool rather than market factors. Urban farms relying on labor pools that include volunteers, youth, and job trainees are challenged to provide adequate training, and they require a higher level of supervision in order to maintain productivity and motivation in unfavorable conditions. However, recognizing the benefits of youth and community engagement, farms with large numbers of youth and trainees may choose not to use production practices leading to optimal efficiency and instead emphasize the use of hand labor, allowing the participation of all ages and skill levels.

Urban farmers have found unique avenues to sourcing farming information to enhance the productivity of urban farms. Farmers recognize the benefit of farming information to improve their production, but some struggle to locate these resources through traditional agricultural training programs. Master gardener programs typically focus on noncommercial agricultural production, and university cooperative extension employees focused on urban agriculture are nonexistent in many urban areas. However, to fill this gap in information sourcing, strong grassroot support networks often exist, (e.g., the listservs for the Association for Urban Agriculture or Refugee Agricultural Partnership Program), with experienced urban grower networks serving as mentors.

Regulatory and legal issues   The productivity of urban agriculture is directly impacted by municipal, county, and state government regulations. Zoning and regulatory requirements may dictate the scope of farm infrastructure, including greenhouses, hoop houses, tool storage sheds, and washing and packing sheds, as well as other production factors such as working hours and production techniques (Pfeiffer, Silva, and Colquhoun 2014; Mukherji and Morales 2010). Lack of tool storage or greenhouse facilities may limit the crops that can be grown or the scale of an operation. In some cases, particularly for livestock and value-added products, regulations may dictate what products can and cannot be sold or produced on the farm, thus impacting potential productivity.

Regulatory issues related to water access can significantly impact the production potential and expansion of urban farms. On urban farms, water is essential not only for the irrigation and production of crops but also for ensuring safe food handling (i.e., proper hand washing and sanitation) and proper harvest and postharvest practices. Urban farm sites may have informal access to water, but legally obtaining water for agricultural activities may be problematic. Historically, urban farms have accessed water through hydrants; however, many municipalities limit the use of these sources. As a result, alternative water sources, such as rain catchment systems, are becoming increasingly common (Petersen 2011). Urban farms in water-limited areas, such as the Desert Southwest, are especially challenged with respect to reliable irrigation, but in some cases they have access to community-operated water systems such as acequias (Covert 2012). However, water access through these sources may still be subject to various governing limitations with respect to access and quantity.

Stability

Using an agroecological framework, stability is defined as the degree to which productivity of an agroecosystem is stable in the face of small disturbances caused by the normal fluctuations of climate and other environmental variables. Production innovations may improve the stability of urban agriculture relative to rural agriculture, while land and labor management aspects tend to reduce stability. Enhancing agroecological stability can include many strategies that impact overall productivity, including crop choice, cultivation methods, and pest management strategies. Urban farms have developed innovative production methods that mitigate the challenges facing food production in an urban setting, allowing for increased stability of production and yields.

Infrastructure and production management   The use of farming methods that reliably maintain quality and yield throughout the entire year contributes to production and supply stability. Season extension, intercropping, and succession planting allow for the production of more crops for a longer time, creating more-uniform production over a twelve-month period. Season extension on urban farms follows the same production and market principles applicable for rural growers, often yielding greater volumes of higher-quality produce while reducing pest and disease pressure (Carey et al. 2009). Hoop houses, high and low tunnels, and greenhouses are commonly used to extend the growing season by creating protected, controlled growing environments for crops. Urban farmers often utilize the favorable microclimates along the outside of hoop houses, buildings, and other surrounding structures to further protect plants from the colder temperatures at both the beginning and end of the production season. Farms using season extension techniques harvest produce for longer periods throughout the year, allowing for greater production and income stability over a longer time.

Urban farms sited on multiple lots throughout a city can encounter management challenges that threaten production stability, particularly with respect to effectively managing pests and diseases. As organizations expand, urban farms often procure new land in a patchwork pattern of noncontiguous lots throughout the city. Urban farmers find traveling to these disparate sites challenging and time consuming because of urban traffic, the nonwalkability of busy streets, and the lack of adequate public transportation. With limited visits to a site, farm employees may fail to observe the onset of insect and disease issues, preventing the early identification necessary for effective management, particularly if organic production methods are used. High levels of disease and insect pressure resulting from lack of timely management can cause crop losses and lead to instability in both the yield and quality of produce.

Urban farms have integrated practices to improve stability under the conditions of noncontiguous land access by employing management strategies that minimize the labor requirements of disparate land parcels. Crop selection and planning that places crops with high insect and disease tolerance or low management requirements on the hardest-to-reach parcels can ease the management burden. Well-managed urban farms may experience higher stability than rural farms in terms of pest management, as pests may not be as prevalent in the urban environment as they are in the rural landscape.

Social goals One of the primary challenges to production stability on urban farms is the reliance on a volunteer labor force. As discussed above relative to productivity, unpaid volunteer workers, although well intentioned and often immensely committed, can also be inexperienced and capricious with respect to their work output. As with any organization incorporating a volunteer labor force, many urban farms are confronted with poor performance and nonattendance of unpaid workers because of a perceived lack of obligation, an issue more generally recognized as "the reliability problem" (Pearce 1993). Retention of volunteers and high volunteer turnover have been cited as serious problems for groups dependent on volunteers to accomplish their organizational missions (Skoglund 2006). As the production of quality agricultural products requires a degree of skill and knowledge on the part of the employee, the high turnover rate of volunteer workers compromises the ability of an urban farm to offer products of stable and consistent quality, thus compromising the ability of a farm to grow and maintain contracts and customers. Lack of farming skills in both paid and unpaid employees can further impact production stability through compromising overall production, as described above. However, reduced productivity in terms of crop output may be balanced by nontangible productivity such as job training, community development, youth engagement, and neighborhood green space improvement. Though volunteer labor may reduce the stability of a single farm in the short term, successful programs may increase farming knowledge or neighborhood stability in the long term.

Sustainability

The sustainability of an agroecosystem describes its ability to maintain its productivity through time, in the face of long-term pressures such as prices and land resources. Urban farmers are faced with unique challenges with respect to the sustainability of their agricultural production because of the physical and political environment in which they are located as well as their organizational structure. Initial and continued access to land, compliance with zoning regulations, and income each impact the success of an urban farm (and thus its sustainability).

Land access and tenure   Sufficient land access and tenure present immense challenges to urban farms. Open space that is suitable for agricultural production is at a physical and financial premium in most urban settings with high population densities and limited space. The siting of urban farms is often not driven by the land's ability to grow food, but instead by availability and opportunity. Depending on the particular city, the cost and availability of potential urban-agriculture spaces vary enormously.

Because of financial constraints, many urban farms are sited on inexpensive or free land located in blighted areas of a city. A major mechanism through which urban farmers procure land is donations or cooperative agreements. Often these donations or agreements are written with the option of the owner to rescind the farming agreement. Such land may be subject to redevelopment or transfer of ownership, impacting the longevity of land tenure agreements and reducing sustainability of production operations. If a site on which a farm is located is targeted for redevelopment by either the landowner or the municipality, the farming project will need to relocate operations to another plot, forcing the organization to continually rebuild. Achieving the social goals associated with urban agriculture depends on strong neighborhood ties and community trust of the organization. The periodic moving and rebuilding of gardens resulting from unstable land tenure necessitates the investment of a significant amount of staff time and the development of new community relations, as well as the rebuilding of physical infrastructure such as pack sheds, greenhouses, tool sheds, and soil remediation. In addition to the financial and labor costs of rebuilding, soil quality and the associated ecosystem services are diminished when organizations lacking land tenure are not given the incentive or opportunity to improve soil over the long term.

In attempting to locate farms on sites that allow for long-term tenure, urban agriculture has embraced the creative use of nontraditional spaces, notably industrial buildings, rooftops, and paved areas. Urban farms often adapt existing structures in areas of the city that have been vacated and are available at prices below the market rate of open lots. Vertical farming, the retrofitting of the interiors of vacant buildings to create a year-round growing space in a controlled atmosphere with access to light, water, and temperature control, offers a strategy for urban farms to use these spaces. Growing within or upon these structures allows urban farms to remain in proximity to their target clientele, markets, employees, and volunteers. Locating within blighted areas also allows the organizations to achieve secondary goals, such as revitalizing an ailing community through job creation, social engagement, education, and beautification.

Urban farmers have further expanded their options for longer-term sustainable farm locations with the integration of rooftop farming. Historically, rooftop plantings have consisted of ornamental plants, which have been shown to provide benefits with respect to enhancing structural cooling and decreasing water runoff. However, urban farmers have adapted these spaces to include the production of food, using the sites to supply produce to urban restaurants, retail food markets, or schools (Oxenham and King 2010). Issues with retrofitting existing structures for these purposes limit the expansion of these types of farms, as the original construction may not have allowed for the load-bearing capacity necessary for the production of crops. Additionally, production challenges exist as a result of high winds, high temperatures, shallow growing media, and access to water.

Some urban growers have addressed challenges with sustainability of urban land access by acquiring larger acreages of rural land, allowing a greater scale of food production. The market and community development opportunities of the urban production site combined with the increased produce volumes and stable land tenure provided by rural acreages allow urban-agriculture operations greater access to markets, particularly wholesale and institutional markets. More options for crop diversification and crop rotation additionally lead to greater sustainability of farms through better management and risk mitigation with respect to insect, disease, and weed management.

Policy and regulatory issues Regulatory issues, including those impacting zoning and land use, have important implications for the sustainability of urban farms, affecting nearly all aspects of operations such as siting, production, infrastructure, access to materials, and marketing (Hendrickson and Porth 2012). Zoning that limits the number of employees, land use, noise, or other aspects of farm management places limits on farm production. Furthermore, changes in policy or zoning can challenge the continued operations of urban farms. Though well intentioned, these restrictions limit the scale, efficiency, and environmental and economic sustainability of urban farms. However, regulation concerning urban farm activities changes regularly (Covert and Morales 2014), often to the advantage of urban farms as their benefits become increasingly recognized.

Economics Economic and financial viability are integral to the long-term sustainability of any business, including farming ventures. Urban farms are very diversified with respect to their financial models, often arising from the diversity of their missions. The income streams contributing to an urban farm's budget commonly include typical farm income generated from the sales of bulk produce and revenues from value-added product sales, grants, gifts, and other activities associated with social goals such as educational and community programming. The impact of these different economic models on the sustainability of both the farming and overall organization varies greatly.

Data collected from a qualitative survey of thirty-five urban farms in eight cities across the United States show that approximately one-third of urban farming businesses attempt to derive their sole income from farming activities (Pfeiffer, Silva, and Colquhoun 2014). The business structures of these farms are similar to those of any diversified direct-market vegetable farm; farmers need to be sure that the fixed and variable costs of production do not exceed the gross profits obtained from the sale of crops. The sustainability of the farm as related to the strength of the farm finances is often determined by the ability of the farm to manage costs as well as set and receive appropriate prices. To best achieve this, many of these operations focus on producing high-value crops such as microgreens, herbs, cut flowers, value-added products, or other specialty items sold to high-end restaurants. Microgreens and leafy greens are particularly popular crop choices, as yield and quality can easily be optimized in space-intensive and controlled-environment settings, maximizing the dollars earned per square foot of production. Additionally, these crops have short time requirements with respect to planting and harvest, allowing for a more stable and continuous (often twelve-month) income.

For urban farms that engage in other activities related to organizational goals beyond agricultural production, organizational structures differ in the degree to which finances of farming and nonprofit activities are integrated. For farms that subsidize the nonprofit activities through their production, value-added products such as honey, canned goods, or wheatgrass for juicing positively impact the financial sustainability of the agricultural operations. The additional income that is generated by the sale of these products contributes to offsetting the financial commitments of other organizational goals, such as youth engagement and job training.

The impact of organizational finances on the overall sustainability of urban farms may additionally be influenced by the degree to which an organization's budget relies on gifts and grants. Typically, if an organization receives a gift or grant, the award supports the social mission of the organization. If the production activities are intertwined with a social mission funded through gifts and grants, certain aspects of production—crops chosen, amount planted, and pest and fertility management decisions—may be at risk if an organization experiences a decline in this type of funding. As such, organizations that separate the accounting of agricultural production and social projects may be less at risk for experiencing financial challenges to the sustainability of their farming operations.

Equitability

From the perspective of agroecological system analysis, equitability focuses on several key production and social aspects of urban farms. An evaluation of equitability takes into account how evenly the products of an agroecosystem are distributed among its human beneficiaries. The greater the equitability of the system, the more evenly the agricultural products are shared among the members of the surrounding community. Equitability can also consider the ability of the community to maintain an adequate income, have access to good nutrition, and maintain a satisfactory quality of life. Equitability is often an explicit goal of urban agriculture. Success has been demonstrated in terms of improved access to food, social and cultural networks, and job training.

Improved food access Food production on urban farms has had a positive impact on increasing food access in food-insecure areas (Armstrong 2000; Balmer et al. 2005; Corrigan 2011; Larsen and Gilliland 2009). Evidence demonstrates that urban agriculture increases fruit and vegetable consumption among participants (Brown and Jameton 2000; McCormack et al. 2010). Urban-agriculture food projects evaluated by the Community Food Security Coalition produced 18.7 million pounds of food, with over 726,000 pounds donated for community food consumption (Kobayashi, Tyson, and Abi-Nader

2010). Urban farms may provide food to communities in several ways. Farms located in blighted areas of cities are typically in areas with poor access to healthy, fresh foods; the presence of farm stands featuring the vegetables and fruit grown on urban farms provides convenient access to fresh produce to these communities. In addition, the farm may provide produce directly to volunteers or workers or donate to area food pantries. The development of community-oriented businesses is another strategy some organizations use to deliver produce to community residents, as exemplified by Community Services Unlimited in Los Angeles, California, and Growing Power in Milwaukee, Wisconsin, both of which have developed a café with the goal of providing fresh, local, affordable prepared food and whole produce to community residents.

Community gardens can be an important tool in increasing the equitability of urban agriculture. Individuals willing to farm a plot of land themselves benefit from access to affordable produce and also gain social and cultural benefits (Armstrong 2000; Patel 1996; Teig et al. 2009). Research has demonstrated that people who participate or have family members who participate in community gardens "were 3.5 times more likely to consume fruits and vegetables at least 5 times per day than people without a gardening household member" (Alaimo et al. 2008). These data were supported through other, similar studies (Blair, Giesecke, and Sherman 1991; Corrigan 2011; Teig et al. 2009; Twiss et al. 2003). Depending on the scope, gardens can move beyond a single household producing food for itself to growing food for other community members or local food banks (Balmer et al. 2005; Corrigan 2011). In addition, youth involved in community garden programs report eating more fruits and vegetables and less junk food as a result of their participation (Ober Allen et al. 2008).

Complementary programs Further enhancing their contributions to increasing equitability, urban farms frequently adopt social justice and community engagement goals. Malik Yakini of D-Town Farm, an urban-agriculture initiative of the Detroit Black Community Food Security Network (DBCFSN), summarized this eloquently: "We're not just growing food. We are growing communities as well." Urban farms employ several approaches to growing communities in tandem with growing food. Community gardens, one form of urban farming, have been shown to serve as important spaces for socializing and gathering of the community, and as a catalyst for citizen engagement (Patel 1996; Saldivar-Tanaka and Krasny 2004; Teig et al. 2009). Urban farms with a predominantly social mission often produce a range of programs that encourage community members to participate in several aspects of the food system, from assisting in growing food to harvesting, marketing, preserving, and cooking food that has been grown in the community. These programs improve equitability not only in access to food but also in broader aspects of social capital, education, health, and financial well-being. Plots managed collectively by community members offer a space where neighborhood residents can gather and build relationships as they grow food. Culturally appropriate cooking demonstrations and nutrition education programs are commonly offered as complementary programs. Seasonal meals and harvest dinners are frequently organized to offer an opportunity to further build community through shared food traditions. Many urban-agriculture projects also address equitability by providing education and materials to support neighborhood residents in creating their own raised-bed gardens, thus increasing food access in yards throughout the community.

Job training is another aspect of production that some urban farms embrace to meet the goal of creating a more equitable food system. Unemployment and poverty can limit a family's or individual's access to fresh produce. Increasing the employment rate in a community can further enhance the equitability of food access as well as the broader goal of household financial security. Growing Home, an urban farm in Chicago, has created a transitional job training program that includes not only agricultural and horticultural training, but also job readiness training, comprehensive referral services, job search assistance and placement, and retention services. Homeboy Industries in Los Angeles uses urban agriculture and value-added food processing as a job-training tool to assist former gang members in developing life and job skills. Though urban farming organizations work toward equitability through diverse strategies, it is a central goal for many nonprofit, community-based projects as well as many commercial urban-agriculture ventures.

CONCLUSION

As farms continue to arise in the urban landscape and associated policy is developed to support these efforts, understanding the interrelated factors that contribute to a sustainable, viable, and healthy urban agroecosystem becomes essential. The analysis of the agroecosystem factors related to urban food production is particularly complex because of the diverse metrics defining the productivity of an urban farm, including the social benefits alongside the quantity of food produced. A meaningful evaluation of any urban-agriculture project must account for this diverse range of production outputs.

The availability of resources—production input, personnel, information, and land resources—recurs as a common limitation to urban production throughout the analysis framework. The expansion of successful models of urban farms across the United States will require efforts by communities, local governments, land-grant universities, and cooperative extension personnel to work toward providing the supports to expand availability of the resources necessary for productive, stable, sustainable, and equitable urban agriculture. Examples of both grassroots and organized efforts to provide these resources are increasing—the US Department of Agriculture recently made significant investments in the development of local and regional food systems, including support for farmers markets and local food business enterprises in urban areas (USDA 2014).

As described above, the production of food in urban environments has demonstrated positive impacts on food security, availability, and consumption patterns in local communities. To expand these impacts, further research to optimize production in tandem with achieving environmental and social goals needs to be conducted. The urban production environment creates unique challenges to production as compared to its rural counterparts. Soil resources, the foundation of sustainable agricultural practices, often need extensive rebuilding and remediation to optimize food safety and productivity. Best management practices to improve soil in the space- and resource-limited environments of urban farms must be developed. Urban microclimates and their related impacts on temperature, humidity, insects, and diseases may vary drastically from those experienced in rural environments. As such, crop varieties best suited for these conditions need to be determined, particularly in the face of the intensive production techniques employed by many urban growers. The unique soil conditions of urban farms as well as intensive production require significant soil- and fertility-building measures. Practices that improve the quality and production process for compost and other soil amendments have enormous potential to improve productivity.

The successful development of urban farms is not without challenges. However, with the innovative nature of many urban farmers, urban agriculture is ripe with opportunity. The process of building sustainable, productive, economically viable, and equitable urban-agriculture systems will integrate new methods and approaches to crop and livestock production with existing practices employed by rural farmers. Organizational missions and social resources will be considered together with management practices to optimize production parameters such as crop species selection, fertility and pest management, crop rotation, field design and row spacing, and harvest decisions. With the development of urban farming research and related resources, urban agriculture can continue to become an increasing contributor to local food systems nationwide.

DISCUSSION QUESTIONS

1. Can/should productivity be measured differently for different kinds of urban farming methods? For example, is good productivity different for an urban farm as opposed to a community garden?
2. Given that compost is crucial to any urban farming operation, what achievable options exist to increase the availability and attainability of compost?
3. Examine Conway's definition of agroecosystems. Are "productivity, stability, sustainability, and equitability" a sufficient measure of all urban agroecosystems? Try using them to measure your community.
4. How should we think about volunteer workforces that very often make up the bulk of urban-agriculture operations? Should they be thought of as a stepping-stone to a paid workforce, or a long-term operational investment?
5. Considering an urban-agriculture operation in another chapter in this book, measure the operation by Conway's definition of agroecosystems. Do these four criteria neatly apply?
6. Income from direct sales of food and community services that urban farms provide and income from grants and gifts are two very different economic processes present in urban farming operations. Which of these is the more ideal economic model for urban farms? Is it a hybrid model? Does it depend on the situation?

7. Silva and Pfeiffer frequently cite the unintentional obstacles of governmental zoning and municipal ordinances as significant barriers to the overall success of urban farms. To what extent should government loosen existing restrictions on the success of urban farms while still accounting for other zoning and public health factors?
8. One interesting side benefit of urban agriculture is that these operations often provide job training. How can urban-agriculture operations best facilitate this community benefit while still maximizing farming?

CHAPTER 8

Lessons from "The Bucket Brigade"

The Role of Urban Gardening in Native
American Cultural Continuance

MICHÈLE COMPANION

INTRODUCTION

Cultivation and urban gardening are critical components of the food system around the globe. Food security studies in numerous countries find that urban food production increases resource access, coping mechanisms, and income-generating opportunities for new entrants into a wage economy, internal migrants, and marginalized groups, including female-headed households (Companion 2012). Agricultural activities facilitate psychological survival as well. Companion (2012) finds that women in Mozambique who engage in cultivation report reduced anomie. Gardening allows them to maintain feelings of connection to their rural homes, traditions, and ways of life. The positive benefits of urban agriculture apply to the United States as well. Obeng-Odoom (2013) argues that "urban agriculture empowers rather than limits" (614). Studies of urban gardens, discussed below, have demonstrated positive impacts, including exercise, crime reduction, personal capacity building, community building, and civic engagement.

However, few studies have concentrated on the role of community gardens in cultural continuance. In indigenous cultures, traditional foods, medicines, and rituals are intertwined with health, spirituality, and lifeways. As a result of urbanization and nutrition transition, some cultural disconnects have widened. Flachs (2010) points out that "food and foodways are important keys to cultural identity. . . . Yet, a growing body of research shows that Americans, especially urban, low-income individuals, as well as people of color, have become disconnected from their food" (1). This has led to negative long-term health outcomes. Physiological pathologies are enhanced by the inability to access and consume fresh produce.

This issue is critical for low-income, urban Native Americans. This study will demonstrate the potential of gardening programs to address the structural, sociocultural, and organizational barriers to healthier eating patterns in this community. Programs have been constructed around nutritional education and container gardening, focusing on social empowerment and capacity building. Participants take greater control over their salt and fat intake by growing fresh herbs. The program fosters cultural continuance through several avenues, including facilitation of communication between parents and children and the production of spiritual resources such as ceremonial sage.

BACKGROUND

Nutrition transition is a process associated with modernization and urbanization, which contributes to structural barriers to healthier eating patterns. Nutrition transition occurs when populations shift from being underweight with infectious diseases to being overweight or obese and suffering from nutrition-related noncommunicable diseases (Astrup et al. 2007; Compher 2006). Populations move away from nutritionally dense traditional foods, which have more abundant fiber, minerals, and proteins (Kuhnlein and Receveur 2007), to processed energy-dense convenience foods and sugary beverages (Popkin 2004). There are negative health consequences for the world's poor (Raschke and Cheema 2007), indigenous people (Damman, Eide, and Kuhnlein 2008; Foley 2005), and the Native American population (Companion 2013b, 2008; Gittelsohn et al. 2006) as a result.

Poverty creates additional structural barriers to good health by limiting access to healthful foods (Companion 2013b; Hu et al. 2013). Barriers include prices, retail options, distance to stores, transportation, and availability of food choices. Lack of resources, combined with constricted shopping opportunities in urban "food deserts," promotes the purchase of inexpensive foods that can stretch the quantity of a meal (e.g., potatoes, pasta) (Companion 2013b; Zenk et al. 2005a; 2005b). Lack of or limited access to full-service grocery stores or farmers markets, combined with low availability of transportation, compels families to stock up on nonperishable items (Ard et al. 2007; Companion 2013b; Hawkes 2008; Hsieh 2004; Lake and Townshend 2006). As a result, consumption of processed foods increases, displacing fresh fruit and vegetables from diets. These factors make low socioeconomic status a powerful risk factor for poor health outcomes, including obesity and type II diabetes (Companion 2013b, 2008; Halpern 2007).

In addition to broad social and economic trends, individual health status is also influenced by interactions of cultural forces and social relationships (Hsieh 2004; Larson and Story 2009), which generate food preferences. Frohlich, Corin, and Potvin (2001) emphasize collective lifestyles, which create and reinforce patterns of food consumption. Delormier, Frohlich, and Potvin (2009) expand on this, focusing on food and eating as a social practice rather than an individual behavior. They note that "eating patterns form in relation to other people, alongside everyday activities that take place in family groups, work, and school" (217). Children are central in shaping the food environment, as they have tremendous influence on shopping and preparation patterns (Companion 2013a, 2013b; Damman, Eide, and Kuhnlein 2008; Foley 2005; James 2004; Lake and Townshend 2006; Raschke and Cheema 2007; Stevenson et al. 2007; Wiig and Smith 2008).

Companion (2013a, 2013b) identifies specific sociocultural barriers to healthier food consumption within the Native American community. One of these barriers is family food history, which is essential in constructing taste preferences. Larson and Story (2009) believe that "cultural food patterns influence food consumption in several ways: they dictate what food is eaten, when it is eaten, and how it is prepared" (S65). Frohlich, Corin, and Potvin (2001) reinforce this in their discussion of collective lifestyles, noting that it is the interplay of norms, values, kinship, and social ties that shapes food preferences.

Historical factors and public policy influence diet as well. When these are applied to Native nutrition transition, Foley (2005) finds that colonial foods morph into new traditional foods over time. Colonial foods include all recipes that were created out of food rations provided by a government. During the removal and assimilation periods in the United States (1830–1928), this usually included the same rations provided to soldiers (flour, salt, lard, coffee). From these rations, foods such as bannock and fry bread were created. Bannock and fry bread share the same base ingredients; both are made from white flour, water, salt, and often sugar and are generally fried in lard. Both are high in fat as a result of frying, high in sodium and carbohydrates, and low to completely lacking in micronutrients. In short, these "traditional" foods are hardly healthy.

Foods such as fry bread become identity markers that maintain a sense of shared identity and connection to that larger group. The prevalence of participation in the federal Food Distribution Program on Indian Reservations, which provides commodity food bundles once a month, combined with strong migration patterns to and from cities, reinforces eating patterns and preferences as well. Families have their own recipes for preparing commodity items, such as Spam. Thus, certain recipes help respondents bridge their personal histories and Native identity to their current experience.

As a result, Companion (2013a, 2013b, 2014) finds that consumption patterns are a form of cultural capital. Cultural capital refers to sets of knowledge, skills, and behaviors that enable individuals to embed themselves in specific subcultures and be recognized as members of that group. Cultural capital is also expressed through the explicit rejection of assimilative forces and reaffirmation of indigenous identity through food. As with studies by James (2004) and Hu et al. (2013), certain food consumption and preparation patterns, such as not frying meat and vegetables in lard or having a salad as an entrée (Companion, 2014, 2013a, 2013b), are perceived as conforming to the dominant culture and giving up heritage. These factors create a strong sociocultural disincentive to alter eating patterns within the Native American community.

RATIONALE FOR THE PILOT PROGRAM

Given the barriers to healthier eating, two programs have been developed for urban Native Americans in four large midwestern cities. The first is for children and at least one parent (Companion 2014), and the second is for adults only (Companion 2013a). Program modules focus on skill development and capacity building to empower individuals to take greater control over their food preparation. They link nutrition and food choices to tradition, culture, and spirituality. Studies show that this increases the transmission of cultural knowledge, revitalizes cultural practices, reaffirms a positive collective identity (e.g., "Healthy O'odham People"), and helps establish and reinvigorate social ties (Companion 2008).

The vehicle for nutritional education for both programs is container gardening. This is appropriate for several reasons. First, Companion (2013b, 2014) finds a lack of familiarity with herbs and spices and a heavy reliance on butter, lard, salt, sour cream, and ketchup in daily food preparation. Growing fresh herbs, combined with demonstrations and tasting, increases familiarity and encourages their use. This can be done in urban apartments, addressing issues of ready access, time constraints, safety, and land availability.

Second, workshops on gardening, along with cooking demonstrations and tastings, can be accommodated at any urban Indian Center with a kitchen and meeting space. Holding programs at Indian Centers can help address some organizational barriers to more healthful eating. Many centers are well established in their localities and are aware of the economic, human, and social capital needs of their clientele. Their partnership in program development is essential for both community outreach and construction of culturally sensitive educational components. Center staff reported that the project allowed them to connect with participants in a different context, enabling them to get to know participants and have a better understanding of their personal challenges. One director believed that this would make him better able to serve his community. Thus, programs increase connection between center staff and clientele. They also help introduce health messages into a community that has not always welcomed outside organizations, such as representatives from federal agencies and their messages. Hu et al. (2013) note that many urban minority communities distrust outsiders. Programming through community-based channels, such as Indian Centers, addresses this barrier.

Third, cooking and food represent significant emotional ties to place and history (Companion 2014, 2013a, 2013b). Hu et al. (2013) find that traditional methods of cooking food and the perceived lack of cultural saliency of healthier cooking methods represent sociocultural barriers (71). Container gardening can increase access to food and reintroduce the cultural connections to healthier foods, while strengthening ties to place and history. With the help of Indian Center staff, spirituality and emotional health can be integrated into a broader nutritional platform in individual and tribally specific ways.

METHODS

Four urban Indian Centers in the midwestern portion of the United States hosted the pilot programs. All are located in low-income, inner-city areas and have food pantries. All have clientele who struggle with hunger and nutrition-related health issues. Center staff requested assistance with program development and evaluation to address community needs, resulting in two programs (one for adults only and one for children and parents) that focus on social and personal empowerment.

The educational modules, which include nutrition, health, cooking techniques, and agricultural skills, are constructed around container gardening. All educational modules are developed in collaboration with Indian Center staff to ensure that the program is culturally sensitive and addresses the population's needs. The findings from Companion's (2013b) survey also informed the construction of the modules.

The gardens are designed specifically for indoor, urban environments. Through donations, the centers provide all the necessary materials for participation in the program. Because of plant maturation time, all participants are provided with started plants rather than seedlings. Most plants are about four inches tall at the time of planting. This cuts down on maturation time but still allows participants to have the educational components related to plant biology. The buckets always contain tomatoes, two kinds of chilis, cilantro, and chives. Other contents vary, based on preference and availability. Windowsill herb gardens containing basil, dill, parsley, sweet marjoram, and cilantro are also provided.

A core program goal, to address sociocultural barriers to healthier eating, is sustainable behavioral modification, making demonstrations and tastings essential components of the programs. Therefore, all foods used in the demonstrations are shaped by local structural barriers. The foods used are easily available in urban food deserts on a low-income budget, with special attention paid to those most likely to be found in food pantry distribution bags. Multiple recipes of commonly consumed foods were generated by local volunteer chefs and nutritionists using both cooked and raw herbs. This enables participants to learn about flavors and cooking techniques and helps reduce the amount of added salt and fats.

Groups meet for three to four hours on Saturdays for eight to ten weeks. Local nutrition and fitness experts, chefs, and gardeners conduct the educational components. A flier was included in food pantry bags and posters were placed around the centers to advertise the program. Participants were signed up on a first-come, first-served basis. In the child/parent program, forty-two children, ages seven to eleven, from thirty households participated. In the adult-only program, twenty-seven men and nineteen women participated.

Initial surveys gather demographic information, reasons for participation, and seven-day food recall information (see table 7). Surveys take forty-five to fifty minutes to administer. To monitor and evaluate the impact of the program, interviews are conducted at the five-week mark, the close of the program, and six months after the program conclusion. Questions are posed in a semistructured format. Interviews average thirty-five minutes. The midterm survey focuses on overall impressions of the program, the utility and accessibility of educational material, interest in and engagement with the educational modules, quality of instructors, and adoption of course objectives. The closing interview focuses on the attainment of program objectives. The follow-up survey focuses on sustainability of behaviors learned in the course and the continued use of information provided. All interviews were conducted and transcribed by the author. Major themes were identified using the inductive grounded theory of Corbin and Strauss (2008).

FINDINGS AND DISCUSSION

In the baseline, being disconnected from traditional spirituality emerges as a primary motivation for enrolling for 61 percent of participants in the adult-only program and 69.7 percent of household heads in the child/parent program (see table 8). Respondents also report feeling separated from living things and nature in the "concrete jungle" (60.9 percent adult only, 56.7 percent child/

Table 7. Participant household characteristics

	Men AO* (n = 27)	Women AO (n = 19)	Household CP (n = 30)
Employment status			
Currently unemployed	14.8%	57.9%	57%
Underemployed (job does not meet skill sets)	81.5%	42.1%	40%
Meaningfully employed (paid well with benefits)	3.7%	0	3%
Food access			
Live within 2 blocks of full-service grocery store	0	0	0
Live within 1 mile of full-service grocery store	3.7%	0	0
Own a car	11%	5.3%	0
Rely on local shops for most food	70.5%	84.2%	20%
Food constraints			
Ever used a food pantry	100%	100%	100%
Ever used Indian Center pantry	100%	100%	100%
Currently receiving food stamps	48.1%	94.7%	67%
Household experienced food shortage last month	100%	100%	100%
Skipped a meal in last week	70.4%	89.5%	87%
Reduced portion size in last week due to lack of food	70.4%	100%	100%

*AO = Adult Only; CP = Child/Parent.

Table 8. Baseline survey results

	Men AO* (n = 27)	Women AO (n = 19)	Household CP (n = 30)
Reason for attending			
Money/constraints on food	48.1%	73.7%	86.7%
Health issues	88.9%	94.7%	93.3%
Disconnection from spirituality/nature	77.8%	36.8%	56.7%
Entertainment/something to do	0	31.6%	36.7%
Home environment/dreariness of where they live	18.5%	47.4%	20.0%
What are you hoping to get out of this class?			
Skills/knowledge	92.6%	89.5%	100%
Food	59.2%	78.9%	70.0%
Connection to past/culture	88.9%	42.1%	90.0%
Connection to other Native Americans	25.9%	89.5%	93.3%
Connection to self	81.5%	57.9%	46.7%
Confidence	14.8%	26.3%	30%
Importance of foods for spirituality			
High	85.2%	100%	100%
Moderate	11.1%	0	0
Low	0	0	0
Current use of foods to link to spirituality			
Daily	0	0	0
Weekly	0	0	0
Monthly	0	0	0
Past year	3.7%	31.6%	53.3%

*AO = Adult Only; CP = Child/Parent.

parent), leaving some homesick and others depressed. This is intertwined with addictions and struggles to maintain sobriety for both groups.

The need for greater spiritual and natural connections is reaffirmed when participants are asked what they wish to get out of the course. While practical outcomes such as skills and knowledge accumulation (91.3 percent adult only, 100 percent child/parent) and food (67.4 percent adult only, 70 percent child/parent) are cited, the concept of embracing spirituality by reconnecting to their past and culture (69.6 percent adult only, 90 percent child/parent) and to themselves (71.7 percent adult only, 46.7 percent child/parent) is also strong. An essential component of this is being able to meet other Native Americans. Approximately 52 percent of the adult-only and 93 percent of child/parent respondents want a core community constructed around other Native people.

Given these responses, a second interview was conducted to identify ways to address spiritual needs through gardening. "Smudging" (bathing in the smoke of burning sage or other medicinal plants) or ritual purification was identified as an important cultural and personal act. Despite its importance, only 15 percent had smudged recently. Respondents report that access to "clean" resources (the right kinds of sage, harvested and handled in a culturally appropriate way) is a primary constraint.

Program participants submitted a list of plants that they would like to grow to help them spiritually. The programs opted to experiment with a second bucket containing beans, heirloom corn, and pattypan squash (the "three sisters"), as well as two types of sage. This combination of plants is based on those historically used by numerous Native American tribes to increase yield in small areas with minimal agricultural inputs. The beans (nitrogen fixing) are trained to climb the corn stalks (nitrogen depleting). Squash is trained to climb out of the bucket and across the floor.

THE IMPACT OF PROGRAM PARTICIPATION

Gardening and cooking together opened new channels of communication within families, helping to address some of the sociocultural barriers. Adult-only participants were provided with suggestions and activity cards to facilitate the transfer of knowledge from the program to their homes. Child/parent participants were required to prepare and cook meals together. All participants reported that the acts of cooking and working with the plants together forged new opportunities for discussions and sharing.

Parents found it easier to discuss their childhoods, food knowledge, cultural knowledge, and memories of family members, rituals, and spiritual events when prompted by the natural course of the growing cycles. One participant remarked, "We were looking at some spots on the tomatoes. It made me think of a funny story my aunt told me about Coyote trying to steal food. I told the kids. They liked Coyote and wanted to learn more about him." Another respondent said, "My kids were real surprised when I came home with those huge buckets. I have a black thumb! But, I told them about what each of the things were in there. When we got to the corn and sage and stuff, I explained about 'the sacreds' and why they are important."

Approximately 73 percent of all participants reported that program participation had a moderate to high impact on their personal sense of spiritual and/or cultural connection. The proportion of participants who engaged in smudging on a weekly basis climbed to approximately 88 percent. This supports Kingsley, Townsend, and Henderson-Wilson (2009) and Hale et al. (2011), who find that urban community gardens impart a sense of connection to nature and traditional cultural history.

Others used cooking to transmit personal and cultural knowledge. More than 92 percent of all participants reported that the program has had a moderate to high impact on their ability to connect food, including commodities, back to their culture and family history. Food is a catalyst for discussions with family about their memories of preparation techniques, unique recipes, or cultural traditions. Despite its importance, however, the majority of respondents (82 percent) are able to do this only once a week because of food access limitations and financial constraints.

A number of those confronting addiction issues believed that the act of caring for the plants, being responsible for them, and seeing the fairly immediate response to their neglect has helped with their journey to sobriety. Several felt that the connection to others who are struggling and are also Native provided incentive to not defect from their obligations. This supports findings demonstrating that gardening promotes environments supportive of healthy behaviors through sustained social interaction (Hale et al. 2011; Kingsley, Townsend, and Henderson-Wilson 2009). Armstrong (2000) believes that gardening is particularly important for health promotion in minority communities, noting that it provides social support and emphasizes informal networks, thereby "encouraging interpersonal, peer-to-peer tactics for promoting change" (325).

These findings also support previous studies that demonstrate the emotional and spiritual healing qualities of gardening (Armstrong 2000; Hanna and Oh 2000; Reuther and Dewar 2005; Teig et al. 2009). Contributing to the positive overall impact, gardens have also been shown to provide a sense of purpose and direction, while furnishing a sense of accomplishment that reinforces personal and social empowerment (Armstrong 2000; Hale et al. 2011; Kingsley, Townsend, and Henderson-Wilson 2009).

An unexpected benefit from program participation was the desire of some participants to engage in larger-scale agriculture. Nine participants applied for space in urban community gardens the summer following the program. Based on their example, an additional six applied for space in the next round. All wanted to produce food and spiritual resources for their families and the community. All mentioned contributing some of their extra food to the Indian Center pantries to help improve food access for other families. Growing and distributing or selling food locally for no or low cost helps reduce some of the structural barriers associated with food access in the area. The increase in civic mindedness and community engagement among those gardeners echoes previous studies that suggest that gardening facilitates value formation, neighborhood attachment and improvement, sense of community, collective efficacy, and increased civic engagement (Flachs 2010; Hanna and Oh 2000; Litt et al. 2011; Reuther and Dewar 2005). Again, this helps reduce sociocultural barriers.

There are numerous other examples of expanded community engagement. Two participants received contiguous plots at a community garden. They are hoping to expand by bringing in a third participant, who also applied for garden space. Their goal is to produce enough extra food to sell at (or to) the Indian Center at a minimal price in order to reduce structural barriers to fresh produce. They also want to produce clean spiritual resources for center members. They plan to use the space as an informal after-school program, where they can teach community children about gardening while getting volunteer labor.

Their example is influencing other center members. To increase interest in the programs, two Indian Centers have featured photo displays of the "Bucket Brigades" and the community gardens. One woman who viewed the photos and read their story stated, "That is a really great idea! I want to go through the course. . . . I want to talk to [one of the gardeners] about maybe working together. I can use the extra corn husks to make dolls and we can sell them there as traditional toys." These results support previous studies that demonstrate that gardens contribute to a community's social capital by improving social networks, organizational capacity, and community pride (Armstrong 2000; Brown and Jameton 2000; Companion 2013a; Hale et al. 2011; Teig et al. 2009).

CONCLUSION

Barriers to healthier eating patterns in low-income, urban Native American populations are complex and intertwined. Structural barriers are related to economic resources and the built environment. These include pricing structures, the existence of grocery stores, distance to those stores, access to transportation, and product choices. The programs discussed above have helped reduce these barriers by increasing food access in the community and at home.

Sociocultural barriers include a network of social relationships and traditional cooking methods and recipes that encompass tribal, colonial, and commodity program foods (Companion 2014, 2013a, 2013b). Perceptions about the cultural saliency of "healthy options" contribute to these barriers. The programs have addressed these barriers by improving interpersonal networks, encouraging civic and community engagement, and facilitating intrafamily communication and reconnections to traditional spirituality and tribal cultures.

Finally, organizational barriers, such as distrust of outsiders, present challenges. This issue was minimized by working with local organizations. Program participants patronized the centers, so a relationship already existed. Collaborating with the staff to create the program and having them participate in the implementation reduced perceptions of outside intervention and improved reception of concepts that were introduced throughout the program.

While this study represents an initial step toward addressing these barriers, the results are promising. Through the container gardening program, the majority of participants reduced their use of salt and fats in food preparation and increased their knowledge and use of herbs and spices (Companion 2014, 2013a). Communication and cultural transmission between family members increased, providing a sense of connectivity to traditional heritage (Companion 2014). Participation created a sense of shared identity and community and provided social support for those struggling with health and addiction issues (Companion 2013a).

These findings have strong implications for health and public policy makers. Funding expansion of Indian Centers so that they can support a demonstration kitchen and creating targeted development grants to provide food, cookware, and other resources for such programs can expand nutritional information and health outreach to an underserved community. Funding for the creation of urban gardens and appropriate zoning laws can lead to more successful capacity-building programs. This will reduce structural barriers to healthier eating by increasing food access.

Collaborating with center staff can help ensure that program modules are culturally sensitive and relevant. This increases the likelihood of overcoming some sociocultural and organizational barriers and attaining sustainable, positive outcomes. More research needs to be conducted on this type of program development and long-term community impacts. It is possible that community garden programs can help mitigate some of the negative health outcomes associated with low-income diets among urban Native Americans.

DISCUSSION QUESTIONS

1. Companion defines nutrition transition as "a process associated with modernization and urbanization, which contributes to structural barriers to healthier eating patterns." Looking beyond this case study, have we seen nutrition transition in history before? Explore how and why nutrition transition has been a salient issue among Native American communities.
2. Companion notes the increased efficacy of help coming from native Indian Centers as opposed to external help that may have bred mistrust. Is there still an argument or a place for external aid regarding food in the Native American community?

3. Is there a compelling case for urban agriculture as a form of therapy for mental health and substance abuse in the Native American community? How effectively can urban agriculture confront this issue? Discuss.
4. Both the adult-only and child/parent case studies yielded positive results. Many program participants expressed desire to continue the next year. Could similar programs, implemented in the same way, be sustainable over time in urban Native communities?
5. One striking, unique quality of these case studies was the use of personal property and windowsills for growing. What effects, if any, did this have on the ultimate impacts of the case studies?
6. What key features and successes of these case studies could urban agriculturists from different cultural backgrounds replicate for their communities' benefit? Which benefits and features are unique to or ideal for Native American communities?

CHAPTER 9

Foregrounding Community Building in Community Food Security

A Case Study of the New Brunswick Community Farmers Market and Esperanza Garden

LAURA LAWSON, LUKE DRAKE, AND NURGUL FITZGERALD

INTRODUCTION

Community food system thinking requires attention to the interrelationships that shape the needs, resources, and opportunities within a physical and social context. A comprehensive community food security strategy starts by clarifying the needs and existing resources within a community and developing a suite of strategies—food policy councils, farmers markets, educational programs, urban gardens, and so forth—that will address issues of access, affordability, cultural appropriateness, and ongoing sustainability (Kaufman and Bailkey 2000; Winne 2008; Raja, Born, and Russell 2008). Given that every community has its own political, socioeconomic, and environmental context, the starting point often involves engaging stakeholders—public agencies, nonprofit service providers, and advocacy (Pothukuchi and Kaufman 2000). In practice, however, developing a multifaceted project can be difficult because of the challenges of communication, negotiation among multiple stakeholders, and the appropriate direction of resources. Particularly when the networks involve institutions and stakeholders who seek to assist a community, the balance between community capacity building and neoliberal or paternalistic engagement requires careful and open discussion of power in decision-making for program development and evolution (Drake 2014; Harris 2009). Even if the implemented project creates new opportunities for food access, its success in practice may still face challenges in garnering the intended participation because of hidden personal and household costs, such as adding a farmers market to a household's already-complex shopping routine, lack of time and experience to participate in urban gardening, or discouragement because of a program's inattention to cultural practices and norms. A community-appropriate approach draws on a network of stakeholders, must be attuned to the needs and practices of the intended community, and must continually reflect on its own decision-making process to ensure balance and responsiveness.

Community food systems are a topic of research and engagement that draws university interests from many disciplines. As with other community-university engagement efforts, universities may become involved in a local community food security endeavor for a range of reasons: for research purposes, as a partner required for a grant, as an avenue for community outreach, or as an anchor institution within a community that seeks to play an active role in its improvement (Sorensen and Lawson 2012; Reardon 1996). While community-university partnership models emphasize participatory approaches, participation has varied definitions that may range from token or symbolic participation to real power vested in local groups (Arnstein 1969; Ostrander 2004; Mayfield and Lucas 2000; Kellogg Commission 1999). Academics may have considerable knowledge to share, but they may not be able to transfer knowledge into action or into the tangible forms of assistance often desired by the affected community (Stoecker 1999). In the context of a community food security project, universities hold expertise in a range of areas that might benefit a program; for instance, social scientists might assist with community organizing and natural scientists might advise on organic horticulture techniques. The question remains, however, whether this is simply technical assistance or a community partnership.

This essay explores these themes through the case of the New Brunswick Community Farmers Market, a project that was initially driven by multiple departments within Rutgers University, the Johnson & Johnson corporation, and nonprofit organizations in the city of New Brunswick, New Jersey.[1] To understand the dynamics of such a project, we find it useful to examine it as an iterative and evolving process rather than a set of sequential steps known in advance. Instead of simply constructing a market pavilion and expecting the program to emerge "naturally," much work went into building organizational capacity. By cultivating relationships with community leaders and developing ways to assess and discuss performance over time, market staff were able to facilitate an evolving program oriented to long-term benefit. Although each group involved certainly stands to gain—whether through publicity, research, outreach, new customers, or improved food access—the acknowledgment of shared benefits across participants has led to a representation of the project as community owned.

DEVELOPING THE MARKET

The city of New Brunswick is roughly five and a half square miles, with a population of approximately fifty-six thousand people, and is located in an urban context surrounded by other municipalities. Established as a colonial town, New Brunswick has transitioned from an economic base in agriculture and trade, to industry, to medical and pharmaceutical complexes. It is also home to Rutgers University, a Research One university and the state's land-grant institution.[2] Like many northeastern postindustrial cities, New Brunswick has experienced demographic change, disinvestment and aging infrastructure, and a shifting local economy. Although located in the city since 1885, the Johnson & Johnson Corporation considered relocating to a suburban location in the late 1970s, in part because of the city's poor physical and economic condition. Instead, however, the corporation decided to build its world headquarters in New Brunswick and establish a public-private partnership to help revitalize the city through block grants and other forms of support (Johnson & Johnson, n.d.). These two institutions—Johnson & Johnson and Rutgers—have many reasons to be involved in local community development.

While located in one of the wealthiest metropolitan areas in the United States, New Brunswick shows striking income inequality within its borders. In some parts of the city, the median annual household income is over $70,000, but citywide nearly 28 percent of residents live below the poverty line (US Census Bureau 2010). Forty percent of city residents are food insecure (i.e., often or sometimes not having enough food or the kinds of foods they want to eat), and in some segments of the city 51.8 percent of residents are food insecure (Fitzgerald 2010; Weiner, MacKinnon, and Puniello 2011). Another health concern is obesity, with 49 percent of boys and 44 percent of girls overweight or obese (Lloyd et al. 2010).

As of 2008, desire to address local food insecurity had been fomenting for many years, including an unrealized proposal to start a downtown farmers market; however, a chance late-summer meeting at the Rutgers Gardens Farmers Market provided the catalyst for action. Just a year prior, the Rutgers Gardens Farmers Market had been established, in part as a way to showcase the commitment of the Rutgers School of Environmental and Biological Sciences (SEBS) to local agriculture and agricultural education. This market quickly became a success, attracting a primarily suburban and university clientele and serving an increasingly varied vendor base. Unfortunately, however, the location was reachable only by car and was distant from the downtown communities of New Brunswick. Amid the stalls and food, Rutgers Gardens director Bruce Crawford facilitated a conversation about the need for a market to serve downtown residents, involving Professor Mark Robson, dean of Agricultural Programs and then-director of the New Jersey Agricultural Experiment Station; Professor William Hallman, who also directed the Rutgers Food Policy Institute; and Colleen Goggins, at that time the worldwide chairperson of Consumer Health Care for Johnson & Johnson. According to Hallman, the group went into action quickly, drafting a proposal to Johnson & Johnson to help fund the New Brunswick Community Farmers Market. By late November, the proposal was accepted with the idea to open a new market by the following summer. As Goggins noted, "Everyone was on board right away—there wasn't any quibbling; it was just a matter of problem solving."

The pieces were in place to start planning the project: expertise and space resources from the university, outreach through Rutgers Cooperative Extension, existing partnerships with local nonprofit organizations, and funding

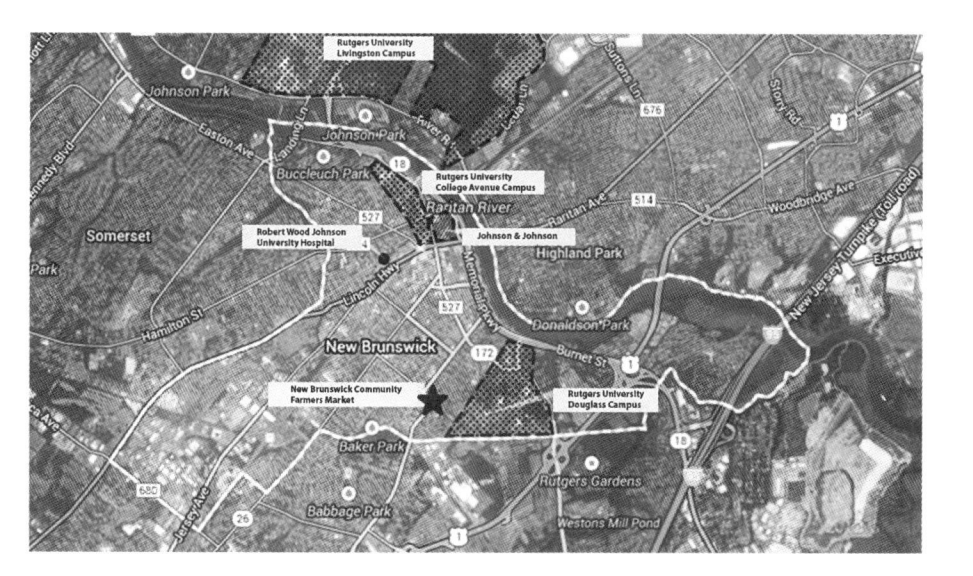

Figure 9. Context map. Illustration by Han Yan.

from Johnson & Johnson.[3] An Executive Committee, made up of Rutgers and Johnson & Johnson leadership, developed the market's mission, which included the following goals:

1. Increase access and consumption of fresh fruits and vegetables, especially by low-income, food-insecure residents of New Brunswick.
2. Improve food intake–related behaviors by addressing barriers.
3. Serve as a catalyst for neighborhood revitalization, creating positive opportunities for Rutgers students and faculty to interact with the community.
4. Offer a convenient venue for education and outreach on a variety of topics related to food, nutrition, gardening, and health.

The team went to work to find an appropriate site (see fig. 9). Earlier proposals to locate a farmers market in the downtown area had faced obstacles about adequate parking space and loading/unloading capacities for farmers. Utilizing an opportunity at hand, the team identified a site owned by Rutgers University that was adjacent to a community with the largest food access need in the city. The site had been occupied by a dilapidated industrial building that had recently been demolished.[4] The site was bordered by the SEBS campus, an established African American community, and a quickly evolving Latino community that included many people from the state of Oaxaca, Mexico. Because it was a university site, development could be handled through the university's facilities offices, thereby minimizing possible bureaucratic obstacles. The city administration offered to assist through attending to temporary parking and traffic concerns. After the initial setup of temporary tent structures, a permanent wooden pavilion was built that provided protection from the rain and increased the market's exposure from the nearby streets.

From initial conception, the organizers acknowledged the tension between institutional power and community ownership of the project. Although the collective capacities of Rutgers, Johnson & Johnson, and the city administration were necessary to successfully establish the program, there was concern that local residents regarded these groups as out of touch with local realities and values. While everyone agreed on the need for food access, effort still had to be made to overcome a community perception that the project framed them as research subjects rather than as project partners. To address this, the organizational structure expanded to include both the Executive Committee, made up of representatives of institutional partners and tasked to direct day-to-day administration, and an Advisory Committee, made up of leaders and representatives of churches, schools, and nonprofit organizations and intended to provide a voice for the community in determining the market's long-term directions. The team was also sensitive to sense of ownership and proposed to name the project the New Brunswick Community Farmers Market (fig. 10), avoiding any reference to Rutgers or Johnson & Johnson so that it did not appear to be controlled by either of these locally dominating entities. Although the funding and administration were directed by these institutions, these key actors recognized that community motivation was needed in order to have meaningful engagement.

Communication of intent and transparent process were essential, as illustrated in the negotiation over assessment and data collection associated with the project. The academic and corporate interests projected the need to develop an assessment strategy to ascertain preconditions and possible barriers to participation and to systematically assess impact after implementation. Some

Figure 10. Farmers market, 2009. Photograph by Nurgul Fitzgerald.

community leaders, however, were skeptical of expending time and energy on assessment and were concerned that it might hinder the focus on tangible actions and bring little value to the community. Why, they asked, should the community invest in assessment? Nurgul Fitzgerald, an associate professor who coordinated the nutrition education and community assessment activities of the farmers market, recalled a participant at the first Advisory Committee meeting saying, "We will be lucky if we can survey fifty residents." The project team navigated the different attitudes about assessment by developing a three-pronged approach involving in-market, residential (door-to-door), and community organization–based surveys. The Advisory Committee supported the endeavor and facilitated access to residents through its extended community networks. During the first three weeks of the market season, 552 people participated in the initial surveys, and another 435 surveys were conducted at the end of the first season. Additional customer and vendor satisfaction surveys took place at the market throughout the season. Analysis of the survey data was used to shape market operations, including market days, hours, what products to sell, and prices. The data also helped direct the nutrition education and other services at the market by assessing nutrition knowledge, food intake, and perceived barriers to food access. In addition, the data made it possible to establish a baseline for monitoring the impact of the farmers market.

While the Advisory Committee consisted of committed individuals bridging the goals of the project to their various institutions, and the institutions themselves engaged many people in the project, it was the direct staff members who became essential to enabling communication among engaged groups and community building. After an initial falter in hiring a market manager from outside the community, Rutgers hired Jaymie Santiago, a Spanish-speaking New Brunswick resident who was already participating in the community nutrition education efforts at Rutgers, particularly the Supplemental Nutrition Assistance Program Education (SNAP-Ed). Attuned to the daily life in this community, Santiago sought to engage residents through existing networks and his active presence on the site and in the community. He reached out to houses of worship, asking church leaders to include market activities in their announcements and to encourage their parishioners to take part in them. He also reached out to schools, day care centers, and service providers. Although he was a staff member of Rutgers, he approached the market from a community perspective, "not advertising it as a Rutgers project but advertising it as a community farmers market. . . . It was shockingly easy to get the buy-in when you didn't go there with an agenda, which most people feel the university comes with or the city comes with. Just, it's just a farmers market, we're just selling food. And people were very happy to welcome it, to share in it." The market manager communicated with residents on a daily basis, met with the Executive Committee on a monthly basis, and reported to the Advisory Committee on a quarterly basis. This time scale reflects the fact that although funding and administrative capacity were top down, constant interaction with community members and consistent reporting to community leaders influenced program directions.

Even with a strong organizational structure in place, the project would not succeed without sustained resident participation. From the residents' perspective, what would make this market useful and a part of their household food purchasing routine? Because many of the intended participants had access to federal assistance programs, the market staff worked with vendors so that

Figure 11. Nutrition information table, 2011. Photograph by Laura Lawson.

they understood and accepted payments through programs like SNAP-Ed, the Special Supplemental Nutrition Program for Women, Infants, and Children (WIC), and the Senior Farmers Market Nutrition Program. "Market bucks" and coupons were distributed to residents through community organizations and local programs. Value was added to the federal assistance program vouchers; for example, ten dollars in vouchers purchased fifteen dollars' worth of produce at the market. Participants in the nearby Cancer Institute of New Jersey's "Body and Soul" program received the market bucks and were encouraged to visit the market. The market also became a focal point for other desired health activities, including on-site provision of nutrition information (fig. 11) and scheduled visits by a health-screening bus from a local hospital.

It was not only a matter of keeping the shoppers coming—it was also essential to establish sustained participation from the farmers and producers who would sell at the market. Extension Agent Bill Hlubik from Middlesex County played an important role in identifying and recruiting farmers to consider

Foregrounding Community Building   149

selling at the market. Because this was a new market in a primarily low-income context, the organizers sought measures to minimize the farmers' possible financial risks. The initial proposal specified a buyback program that would guarantee a sufficient level of sales to the farmer. This was accomplished through a Johnson & Johnson block grant that enabled Elijah's Promise, a local nonprofit food assistance organization, to buy food from the market each week. This win-win strategy not only ensured additional sales for the farmers but also allowed Elijah's Promise to select desired produce rather than ending up with unpredictable supplies of leftover (unsold) produce, as is often the norm for emergency food aid organizations. Nonetheless, there was a degree of vendor attrition. In the first year, there were three farmers, but that proved to be too much for the client base to support, and the farmers felt they were unfairly competing against each other for a limited market. One of the initial farmers—Pop's Farm—persisted and adjusted to the particulars of this market. Initially, the farmer came with standard produce catering to an Anglo-American audience. When some of the local participants asked about herbs and peppers they used in traditional Oaxacan cuisine, the farmer began growing and selling those crops. Pop's Farm was further supported when he expanded his sales to a produce and flower stand on the Johnson & Johnson campus, which provided him with further sales.

MARKET AS CATALYST FOR ADDITIONAL PROGRAMS

The New Brunswick Community Farmers Market provided a foundation for additional food access endeavors. Through market surveys and informal conversations, local residents indicated that they also wanted a place to grow their own food.[5] The market manager began the process of creating a garden, reaching out to the Rutgers Department of Landscape Architecture for design assistance. Several undergraduate students, under direction of faculty, measured the site, discussed the program with Santiago, presented alternatives, and ultimately developed a site design. Fortified with design drawings, the future gardeners took the project from design to construction. Santiago had lumber and fencing supplies brought to the site, and the gardeners built raised beds, compost bins, and fencing. Additionally, a metalworker provided materials at cost and helped the gardeners build an arbor over the garden's

Figure 12. Jardín de Esperanza, 2013. Photograph by Laura Lawson.

entryway. The gardeners named it Jardín de Esperanza (Garden of Hope, fig. 12). In the first year, forty families worked in the garden, and since then, it has continued to evolve and expand.

Once the community garden was established, the market manager noticed that many of the gardeners were growing marigolds. Although a native Spanish speaker, Santiago was Puerto Rican and was not familiar with the Mexican Day of the Dead traditions. Gardeners explained the cultural significance of the holiday and the role marigolds played in it. Since the market's approach was community based, he worked with the gardeners to start a marigold income-generation project that complemented another preexisting marigold project on a separate part of campus.[6] The project emerged as a way to express cultural practices while also earning income to buy materials for the garden. Santiago said, "We decided at the end of the year when we harvested [the marigolds], we put on a small demonstration to show people what the Day of the Dead meant. Giving them that cultural connection. Giving the

Foregrounding Community Building   151

Figure 13. Community event featuring Oaxacan dance, 2011. Photograph by Laura Lawson.

gardeners the sense of pride that we took interest in their traditions, and at the same time fund the community garden on a year-to-year basis" (fig. 13). Concern about frost damage led to the decision to move to greenhouses on campus and then to the development of a hoop house at the market site to protect the marigolds and extend the growing season for garden vegetables. Research and technical assistance in constructing the hoop house involved a community development course taught by Bloustein School of Planning and Public Policy professor Kathe Newman (Capece, Cassidy, and Sarsycki 2012).

It is important to note that the project has continued to grow and evolve. Part of this evolution has resulted from changes in staff, leading to new relationships and opportunities. In 2013, Jaymie Santiago took a position with New Brunswick Tomorrow, a nonprofit devoted to social and economic revitalization. The next market manager was Paul Helms, who had previously worked at Elijah's Promise and was active in the New Brunswick Community Garden Coalition. Building on the farmers market's success, he started a second downtown farmers market. This expansion increased the customer base for both markets, and more farmers and producers became involved, including Elijah's Promise's community-supported bakery business. The original site also continued to evolve, largely through the efforts of Juan Rodriguez, graduate student and garden coordinator for Jardín de Esperanza, and Matthew Smith, an undergraduate majoring in agriculture and food systems. These two went to work to add a children's garden, expand use of an old greenhouse, and build a second hoop house. Their efforts improved the marigold and plant-start production. From 2014 to 2016, the leadership changed to Sarah Dixon, formerly the coordinator of the New Jersey Food Corps. Under her direction, the program continued with site development to include water catchment off the farmers market structure, a chicken coop, and a proposed demonstration kitchen, as well as initiation of planning for a third farmers market site in the city.

LESSONS LEARNED

This array of food- and garden-related activities emerged from a desire by Rutgers, Johnson & Johnson, the city administration of New Brunswick, and other local activists and organizations to improve residents' access to fresh, healthy, affordable, and locally grown produce. Key actors worked together to establish a farmers market in a neighborhood with the largest need. The market site eventually came to support a community garden, entrepreneurial agriculture through social enterprise, nutrition education, and a children's garden. The success of the first site has led to a second market, with another on its way. In terms of lessons learned from the New Brunswick Community Farmers Market, it is clear that both an institutionally driven, collaborative approach and a commitment to community engagement and relationship building made it happen. This evaluation reveals three key lessons learned from the New Brunswick Community Farmers Market (table 9).

First, a litmus test for a successful collaborative effort may be in the conscious acknowledgment of varied but equally significant benefits sought by the various involved parties—academics, corporate sponsors, city administration, community organizations, and residents. While providing nutritious food and nutrition education is a shared goal, each active participant has other goals that shape the resources, opportunities, and evolution of the program. The effort

Table 9. Goals and resources

	Goal / benefit	Resources / what they brought to the table
Rutgers University	• Outreach in nutrition and health as appropriate for Extension services • Applied research • Hands-on learning opportunity for students	• Land and water • Technical assistance • Content expertise: grant management, agriculture, nutrition, and evaluation • Existing programs: SNAP-Ed, WIC, outreach, nutrition and health education • Supervision / project manager • Consumers to market
Johnson & Johnson	• Community commitment	• Funding • Coordination between farmers market and Elijah's Promise
City of New Brunswick	• Address need for food access • Neighborhood revitalization	• General support • Removed obstacles and provided support for downtown farmers market
Farmers	• Direct market • Guaranteed sales	• Added herbs desired by shoppers
Residents	• Improved access to healthy food • Improved access to nutrition and health information • Improved opportunities for community gardening • Opportunity to voice their food-related issues • Opportunity to influence their own environment • Economic opportunity to participate as a vendor	• Feedback that shaped the project • Consumers to keep the market alive • Development of garden and children's garden
Elijah's Promise	• Buy food they actually need • Job training / sales at market	• Quality food access for participants in their programs • Becoming a vendor at the downtown market

integrates into the mission the strengths of the institutions and the needs of the community. For Johnson & Johnson, it reflects well on its corporate commitment to the local community. Johnson & Johnson administrators are also able to synergize their support so that it has bigger impact, such as linking farmer sales guarantees with providing appropriate food for Elijah's Promise. The market and its associated nutrition outreach are fitting expressions of Rutgers University's commitment to engaged scholarship and outreach. In particular, the School of Environmental and Biological Sciences, growing out of the Agricultural College, has been in the process of reconstructing its agricultural education in the twenty-first century to include urban agriculture and food access. In addition, nutrition and health are focal areas for the university and a popular concentration field for students in an array of disciplines. Having a farmers market and nutrition education programs in the adjacent community provides not only an outlet for applied education and research but also a confirmation of the important impact of such programs. It supports engaged research opportunities for multiple faculty members and students working in the area of food, nutrition, and urban agriculture (Aminyar and Fitzgerald 2010; Fitzgerald, Czarnecki, and Hallman 2010 ; Fitzgerald and Shah 2010; Spalding et al. 2012). It is also important to shape the program to address the needs of the farmers, who cannot be altruistic but must evaluate their continued involvement based on efficiency, sales, and prospects for future opportunities. And most importantly, the project can succeed only if local residents choose to participate. What is sold, payment options, and hours of operation must align with resident needs. Participation may also mean expanding programming to address other desires, such as community gardens, entrepreneurial agriculture, and children's gardening and education.

Second, an atmosphere of collaboration is created when no one institution tries to "own" the program. Each participating organization invests resources, but the results are bigger than the individual parts because success is tied to community participation rather than solely the institutions' measures of success. An administrative structure, made up of the Executive and Advisory Committees, was put in place that emphasized that community stakeholders would bear serious influence on the long-term direction of the project. Market managers were accessible at the site and worked closely with residents to run programs effectively and to establish new programs.

Third, the continual feedback and appropriately framed data collection provided assessment tools that fed directly into improving the market and expanding programs as appropriate to community interest. Inclusion of assessment in an acceptable manner enabled scholarly research as well as a feedback mechanism to gather community reactions. While assessment can often seem like an "academic" evaluation of success or failure, in this case the surveys and advisory councils identified new needs and opportunities and helped the market evolve. This approach led to not only participation but also an evolving program that identified additional opportunities—community gardening, social enterprise, a children's garden, and much more.

In conclusion, the usefulness of this case study is not in highlighting a successful program but in showing the careful planning and ongoing processes needed to cultivate the required relationships and avenues of communication. The criticism that many community food access projects engage a neoliberal frame must be taken seriously and countered with an intention for shared leadership, active communication, and willingness to evolve based on community residents' needs and desires. Growing out of concern about community food access and a commitment to a community-engaged process, what was once an underutilized property in New Brunswick, New Jersey, became an active farmers market, community garden, children's garden, and site for nutrition education. This did not happen overnight, nor did it happen simply because influential local institutions—the university and a corporate sponsor—decided to make it so. It was shaped by the context and the community so that it would be relevant and therefore be used and successful. Next year and in five years, the market and its associated programs will undoubtedly evolve and change just as the partners' and community's needs change. The lessons learned from this case study confirm that local food security requires not only attention to food access but also to communication, networking, evaluation, and willingness to adapt and change.

NOTES

1. The authors gratefully acknowledge encouragement given by the many people involved with this project and who willingly participated

through interviews and sharing documentation. This work is supported by the USDA National Institute of Food and Agriculture, Hatch project NJ84105.
2. When the New Brunswick Community Farmers Market was being conceived and developed, Rutgers University and the University of Medicine and Dentistry of New Jersey had not yet merged.
3. Dr. Larry Katz, senior associate director of the New Jersey Agricultural Experiment Station and director of Cooperative Extension, has continued to support this program.
4. In addition to the farmers market, the site also served a school garden program, "Nurture thru Nature," which also receives funding from Johnson & Johnson. See http://ntn.rutgers.edu/.
5. The initial proposal to Johnson & Johnson did include a community garden, but it was not included in the beginning phase.
6. The Marigold Project was started by Teresa Vivar and Anne C. Bellows, project staff of a USDA-funded Green House Project. Local leaders established the nonprofit organization Lazos America Unida to continue the effort.

DISCUSSION QUESTIONS

1. What unique advantages are there for urban agriculturists in teaming with universities and businesses?
2. The authors make a compelling case for the benefits of public-private partnerships in post-industrial cities, sometimes referred to as the "Rust Belt." Could benefits of similar partnerships be applied to developing urban farms outside the Rust Belt?
3. In the end, was the New Brunswick Community Farmers Market more of a community project, or a university project? How is this distinction important?
4. Should urban-agriculture partnerships with universities be a short-term measure to promote urban-agriculture efforts, or should they be a long-term model?
5. To what extent is vendor-versus-vendor competition healthy? Can it become detrimental, and if so, at what point?

6. How much influence do universities have over the agendas of urban agriculturists? If such influences exist, are they beneficial overall?
7. Can such private-public partnerships as in New Brunswick sustain urban farmers markets outside the presence of universities? Explain and elaborate.

CHAPTER 10

Fumbling for Community in a Brooklyn Community Garden

DORY THRASHER

The Brooklyn Community Garden no longer exists.[1] From its founding through its end, the garden was fraught with tensions between racial groups, between newcomers to the neighborhood and those who had been there for decades, and tensions surrounding leadership and decision making in garden governance. This discord is rare in the literature on community gardens. The common refrain is that such gardens serve as warm, generous spaces for activity, bringing together community members for neighborhood care and stewardship. For instance, Payne and Fryman (2001) write: "In diverse neighborhoods, an inclusive community garden program may be one of the few institutions that accurately reflects an area's multicultural identity and works to build a united front to address the whole community's needs" (7).

But the Brooklyn Community Garden was not a place where diverse neighborhood residents came together and defused the racial tensions that arose out of rapid neighborhood change. Instead, the garden focused and reflected those tensions. This study provides a counterpoint to the idea that community gardening inherently builds community. Community organizations and other garden organizers have to consider the potential for conflict, as community gardens are not separate from the communities they serve. They import a neighborhood's existing concerns and give those issues an arena in which to be discussed.

The key to minimizing debilitating conflict is to think of community gardens not as apolitical, but instead as collective endeavors whose quality depends on governance (Freeman 1972). Careful attention thus must be paid to the construction of an appropriate governance structure, with clearly defined guidelines surrounding membership, leadership, and decision making, so that conflicts and tensions that may arise around racial tension or neighborhood tenure can be managed or avoided. An understanding of how a garden might fail makes it more likely that future community gardens will succeed.

In this chapter I offer a review of the literature that suggests that community gardens are places for neighborhood cohesion, as well as an explanation of my ethnographic methods. I then tell the story of the Brooklyn Community Garden's brief life, highlighting the participation of key players and some of the tensions that arose while they tried to build a garden from scratch. I then discuss the way conflicts over race and neighborhood tenure resulted in fraught and unstable garden governance and prevented the garden from becoming the idealized community-building space lauded in the literature. I end by offering some final lessons for those who hope to use a community garden as a tool for neighborhood cohesion and community building, and pose some questions about the standard narrative that has come to define community gardening.

HIGHLIGHTING COMMUNITY

Community gardens are thought of primarily as places to produce food in urban settings (Gottlieb and Joshi 2010, esp. chap. 6; New York City Council 2010; Ackerman 2011). They are also routinely promoted, both in the literature and in policy, as sites of community cohesion and well-being, as Twiss et al. (2003), for example, declare: "Community gardens build and nurture community capacity" (1435).

Community gardens can function as havens for neighborhood residents by providing a connection to nature; they can also be sites of civic participation, as garden organizers necessarily navigate permitting, water, and other concerns of city bureaucracy (Schmelzkopf 1995). Gardens further serve as sites for cultivating democratic practice and civic engagement (Baker 2004; Glover, Shinew, and Parry 2005; Krasny and Tidball 2009). Baker notes that gardening "creat[es] an opportunity for people to dirty their hands, grow their own food, work with their neighbors, and generally transform themselves from consumers of food into 'soil citizens'" (305). Armstrong (2000) writes about how community gardening facilitates community building and organizing and leads to other neighborhood issues being addressed. Her research includes examples of gardeners engaging in a fight to keep a supermarket in the area, coordinating a neighborhood watch, and organizing for physical improvements around the garden site such as sidewalk repair.

A number of authors also focus on intergenerational and intercultural connections within community gardens, pointing out that developing relationships with other gardeners was as important to participants as learning gardening skills and growing food, if not more so, and that gardening together leads to sharing food and stories (Krasny and Doyle 2002; Saul and Curtis 2013). Gardens can provide a way for immigrants and people of similar cultural heritage to connect to their agricultural roots in their new environments, finding familiarity with other people and shared activities (Airriess and Clawson 1994).

GARDEN RESEARCH

I began research on the Brooklyn Community Garden with this understanding of the community-building and democracy-promoting potential of community gardens in mind. The garden was located in a long-established working-class African American and Caribbean neighborhood; the neighborhood was changing as a cohort of young, white, well-educated people seeking affordable rent began to move into the area at a rapid clip.[2] The visible changes in the neighborhood—roti shops and hair-braiding establishments closing while bars and cafés opened—were constantly being discussed on the street, at neighborhood meetings, and on the Internet.[3] While all of this discussion highlighted divisions between races, classes, and ages, and between old and new residents, the new community garden had the potential to mend those fractures. The Brooklyn Community Garden was a space where black people and white people, old people and young people, newcomers and long-standing residents were working on a collaborative project in the very center of the neighborhood. Knowing that the literature touted community gardening as a way to build strong place-based associations, I set out to study this particular garden to see those ideas in practice and to understand how a gardening project could work to build community and push against the tide of divisiveness in a neighborhood undergoing rapid change.

From 2010 to 2012, I participated as both a gardener and an ethnographer. I tended a plot, attended garden and community meetings, and subscribed to the garden listserv. I volunteered my skills where appropriate, which included making fliers advertising community workdays. As a researcher I tried to remain neutral in the face of debates or disagreements, but no doubt my position as a white graduate student, and a person relatively new to the neighborhood, influenced how I was able to participate, and thus how this story is told.

THE STORY OF THE BROOKLYN COMMUNITY GARDEN
Year 1: Groundswell

The Brooklyn Community Association (BCA)[4] is a group that started in the 1980s to revitalize the area's main commercial strip. Over the years the group planted trees, put up street banners, and generally brought order and dignity to a neighborhood that had suffered from disinvestment. In May 2010, the association initiated a cleanup of a vacant lot that had long been used for illegal dumping. Members put a padlock on the fence to prevent further trash disposal and began recruiting people to clean out the lot and create a garden. Over one weekend in June, twenty or so neighborhood residents brought shovels and trash bags, cleared the space of garbage, and demarcated areas as garden plots.

After the initial debris removal, the garden began to take shape in a haphazard way. The gates were open on weekends and anyone who walked in and showed interest was offered gardening space by whoever was in the garden. Melissa,[5] who had put the lock on the fence, made keys for those who had participated in the cleanup day, and key holders made further copies for newer joiners. Gardeners hurried to get plants in the ground and get their plots in order, bringing tools to the garden for shared use. In contrast with established community gardens with committees and formal membership, the first year of the garden was somewhat anarchic. "Members" were people who had copied keys from others, "plots" were any areas that people decided to plant, and "open hours" were whenever anyone was in the garden.

When the garden project began, the leadership of the Brooklyn Community Association consisted primarily of African American women in their forties and fifties, all longtime neighborhood residents. It was two middle-aged African American women who owned homes and businesses in the neighborhood—Melissa and Shelley—who initiated the garden. However, the crowd at association meetings and the cleanup day was much more diverse than the leadership of the organization. The new white residents and business owners regularly attended the association's monthly meetings, sitting next to long-standing African American and Caribbean neighborhood residents. Although garden members included a diversity of races, ages, and neighborhood tenure, more than half of the garden members were the young white people who had moved to the neighborhood in the past five years. In this first year of the Brooklyn Community Garden, the garden's ability to draw a diverse crowd into neighborhood participation was regarded positively; at the Brooklyn Community Association meeting in September 2010, Melissa declared that "a great sense of community, not just a garden" was being created.

Some gardeners grew vegetables, others flowers. Some built raised beds, others planted directly in the ground. Cliff, a hospital worker in his late fifties who was originally from the Caribbean, had a plot twice the size of the others where he grew pole beans and collard greens. Sarah, a young, white, full-time garden educator, planted ornamentals in the front of the garden and built a compost bin in the back. People used the space for informal gatherings on weekends and after work. The 2010 season came to an abrupt end when the weather turned bad: a short but heavy hailstorm destroyed most people's plants in November, forcing a quick end to the garden's first year.

Year 2: Whose rules?

Planning In March 2011, Melissa called a meeting to discuss the upcoming gardening season. She reached out via e-mail to people who had been involved in the garden in 2010 and advertised the meeting to the Brooklyn Community Association's listserv and to a neighborhood food justice group. A self-selected group of ten people gathered at a neighborhood bar, most but not all of whom had been involved in the garden the year before. There had not been much contact between the attendees in the four months since the gardening season had ended the previous November. This, plus the inclusion of additional participants, meant that the project's momentum and the group's cohesion had to begin anew.

The first issue discussed was the need to build raised beds—wooden boxes for planting, built on top of the ground and filled with clean soil. One gardener had sent soil samples from the garden out for testing and the results showed that heavy metals were present. This made it unsafe to grow directly in the ground, not just for individuals but also for the garden as a whole, as digging in the contaminated soil would kick up toxic dust that could settle anywhere. Building raised beds would address this problem and also ensure that all plots were similarly sized and efficiently laid out. The group planned a work weekend to collect donated building materials, level the garden's uneven terrain, build raised beds, and lay down flat stones to minimize dust in open areas.

The next topic was membership and its responsibilities. The group decided that after the raised beds were built, plots would be distributed. Those who had been involved the previous year and those participating in the work weekend would be offered the first opportunity to claim gardening space. The question of membership roles and responsibilities was too large to tackle at this meeting, and five or so people volunteered to form a membership committee to consider these issues and draft proposals.

The third topic, a discussion of garden rules, caused the first real rift in the group's growing sense of cohesion. Howard, an African American lawyer in his late thirties who had lived in the area for about seven years (longer than most of the white members, but not decades, like Melissa and some others), proposed that the garden needed a set of rules. He was worried about liability: the toxic soil and the uneven terrain concerned him, and he felt that rules would mitigate responsibility should anything go wrong. Though not everyone was as moved by the liability question, a rules committee seemed reasonable.

As people expressed their interest in forming a rules committee, Melissa insisted that the first rule should be "no dogs in the garden." Not everyone felt that was necessary, particularly Sarah, who often brought her dog into the garden. Melissa responded to this pushback by saying, "Dogs do not belong in the garden. We have to make the garden welcoming and some people are scared of dogs." Joshua, a white man in his midthirties who worked as a website designer, turned to Melissa and replied, "Well, some people are scared of black people, should we not let them in the garden?"

This remark stunned the group into silence. Melissa spoke first: "You did not just go there. You did not just compare black people to dogs." Joshua stood by his statement and added that he thought that Melissa was "bigoted" toward animals. However, many in the group found his comparison to be racist, and this was the start of a serious and prolonged animosity between Joshua and Melissa that was never resolved. The meeting to reassemble a group of dedicated gardeners and plan for the upcoming season successfully established some crucial next steps. However, despite the common goal of moving the garden forward, the disagreement and personal attacks that ended the meeting failed to unite the group and created a rift between two participants that impinged on the cohesion of the entire group.

Building   Plans moved forward. The membership group began meeting. The garden held a work weekend and some raised beds were built and filled with clean soil. One member, whose family owned a garden-supply store, donated garden tools. A different member organized a bulk purchase of seedlings from a local nonprofit. Still, further tensions around race, neighborhood tenure, and leadership arose as spring blossomed.

First, the project to level the uneven terrain of the garden destroyed what remained of the growing areas from the year before, including the double-sized plot that Cliff—the Caribbean man in his late fifties—had been using. Cliff was not present at the workday and was livid when he saw what had been done, partly because he had let things go to seed in place so that they would take root when the weather got warm. For the same reason, he refused to let raised beds be built on top of his plot and continued planting directly in the ground, much to the chagrin of those concerned with minimizing the impact of the contaminated soil. Cliff had been growing vegetables directly in Brooklyn soil for years and was adamant that it was not a problem. Cliff was not a part of the e-mail listserv and he complained directly to Melissa, who then wrote to the e-mail list reporting that Cliff was "very upset," and that the gardeners ought to be "more respectful" of someone who had been living in the neighborhood for decades. Here, Melissa equated angering Cliff with being unwelcoming to longtime neighborhood residents more generally. In the garden, those who were active proponents of raised beds and fresh soil discussed how Cliff's plot made them nervous, but not one of them wanted to be at the center of a generational dispute, and thus, Cliff was left to garden directly in the ground.

Second, the plan to build raised beds throughout the garden and then assign those plots to individuals quickly went awry. Shortly after the work weekend, when only some of the raised beds had been constructed, those who had built them began planting, preemptively claiming beds as "theirs" and not participating in building out the rest of the garden. One garden member, Justine, a white graduate student who had lived in the neighborhood for two years, noted in an e-mail to the garden listserv that this claiming of plots seemed to thwart the membership committee's plan for equitable plot allocation. She suggested that Melissa, who was generally respected as the head of the garden, seemed to be allowing people to start planting rather than protecting the decisions of the membership committee.

This was also the beginning of a passive disagreement over who "ran" the garden. The issue became further contested when, at the beginning of May 2011, Melissa arranged a meeting to discuss the garden's relationship to the Brooklyn Community Association. She invited a select few people to the meeting: Justine and Joshua, both of whom were part of the membership committee; Howard, who had raised the issue of rules at the meeting in March; Shelley, a long-standing officer of the Brooklyn Community Association; and me.[6] Because the garden needed 501(c)(3) status to apply for various small grants, its relationship with the association was important; however, Melissa was becoming concerned that the garden was moving away from the Brooklyn Community Association, as not all people involved in the garden were necessarily involved in the association.

At the meeting Joshua and Justine took the position that the garden was separate from the association, but the relationship was useful. Howard, however, advocated for a much stronger connection, saying, "The BCA put the lock on the fence. The BCA organized the cleanup. The garden is part of BCA."

When Justine asked what that meant going forward, Howard got angry and accused her of trying to "erase history." Howard also brought up the fraught interaction between Joshua and Melissa from earlier in the spring, telling Joshua to his face, "You are a racist and I don't like you." The meeting ended with all agreeing to be civil; very little was decided about the relationship between the Brooklyn Community Association and the garden. Though the gardening season was moving forward, relationships between gardeners were degrading.

Meeting   Later that month, an all-garden meeting was called. The meeting was advertised on the gates of the garden, through e-mails to the garden listserv and the Brooklyn Community Association's listserv, and by word of mouth. The goals of the meeting were to distribute the remaining plots and to allow the membership committee to present their draft of the garden governance structure. Attendees included both those already involved in the garden and new people who hoped to join and were being oriented to the garden for the first time.

The membership committee—the group of five or so people who had been working since March on creating a structure for the garden—discussed the limited availability of plots for the year and other ways that people could be involved in the garden: turning compost, planting in the communal garden area at the front of the lot, and being present for open hours. This sort of participation, they suggested, would increase people's chances of getting plots in 2012.[7] Many people left the meeting when they found out that having a space to garden would be impossible for them that year.

The membership committee then presented the membership contract they had drafted. To their surprise, rather than a discussion about the content of the document, the debate turned to the right they had to impose these rules. The committee members were taken aback; their right derived from their participation in the membership committee, and from their desire to give their time to the administrative task of sorting out how the garden should be run. To be met with vitriol from the group of garden participants was upsetting to them. Chastened, they did not push forward on holding a discussion and voting on the membership structure. The meeting ended with nothing decided; nongovernance had defeated organization. Nevertheless, people continued planting, watering, tending, and harvesting. Compost was turned. The garden space was used for children's art-making workshops, garage sales, and after-work socializing. Though volunteer committees to govern the garden had been established, their legitimacy had been challenged, further exacerbating the tensions between participants.

Infighting The day-to-day enjoyment of the garden did not mean that underlying conflicts had dissipated. In fact, the failure to coalesce was taking new forms. One set of interactions that highlights the ongoing tension involved Tish, an African American woman in her midtwenties who had grown up in the area. In June she sent an e-mail to the garden listserv demanding to know what had happened to the peony bulbs she had planted, claiming that suddenly ornamental plants were in the space she had planted; somebody obviously must have dug up the bulbs. Sarah, whose job as a professional garden educator gave her access to inexpensive plants, had brought the ornamentals to the garden and someone else had planted them, not knowing that Tish had put in bulbs. A few weeks later, Tish wrote again, "Yesterday, Sarah, while planting some ornamental flowers in the common areas I found my peony sprouting under the holly. It was there! . . . I think what happened was either the marker was moved to plant the surrounding shrubs or the peony bulbs were shifted to make room in the soil. Either way I thought I was 'Cliffed' and I was wrong."

This e-mail was a sincere recognition that her bulbs had not been intentionally uprooted, but Tish's feeling of being "Cliffed" points to the initial assumption of antagonism, that someone would casually disregard the work that she had put into the garden, just as the garden group had destroyed Cliff's plot from the year before while leveling and landscaping the garden earlier in the spring.

Later in the season, Tish and Howard suggested hosting film nights in the garden, projecting movies on the wall of the building next door. Though people agreed that it would be a fun activity, Justine voiced some hesitation. She noted that the garden had no electricity to plug in a projector, and furthermore, the garden was on a noisy street that would make it difficult to hear the film. Tish responded to these concerns over e-mail, writing, "I grew up here! I know it's noisy! Don't tell me things like you know better than I do about the avenue!" Even though the garden brought together all sorts of people, to Tish, the distinction between being new to the neighborhood and being a long-standing resident was as salient as ever.

There was no formal end to the 2011 gardening season, no final harvest or celebration. The dwindling cohesion and lack of cooperation in the garden meant that no one was inspired to organize such a thing. Gardeners simply stopped gardening as the weather got colder.

Year 3: Bulldozed (just in time?)

In April 2012, Melissa sent out an e-mail to the garden group with a list of the year's rules for the garden. She asserted that the garden was a project of the Brooklyn Community Association, and that association members had drafted these rules, which included a ban on pets and a requirement that gardeners serve on a committee. The message ended with instructions to e-mail Melissa directly to request a plot. These particular points were all related to issues that Melissa had been concerned with the previous year—the garden's drift away from the community association, her desire to ban dogs from the garden—and had not been resolved by the collective garden leadership. The inclusion of her personal e-mail address was a reclaiming of garden leadership from the committees. Without any protest, many previous garden participants e-mailed to claim garden plots, and the cycle of planting began anew.

However, a week and a half into May 2012, Melissa sent another e-mail titled "Sad News for Garden." She wrote that the owner of the lot was selling to a developer, and that the group had two weeks to vacate the garden. She invited people to come to the garden to relocate plants, signing off, "We understand how upsetting this is, but it is a community project and as a community we will come through this stronger."

DISCUSSION

The Brooklyn Community Garden was bulldozed for development on a lot that had sat empty for thirty years. The neighborhood change that had begun in the years leading up to the garden was characterized primarily by young white people seeking affordable rent; this turned the previously all-black

Brooklyn Community Association into a racially and tenure-diverse group who cared about the neighborhood, and the gardeners were no exception.

The garden facilitated generous, pleasant, and productive neighborhood encounters—for instance, a shop owner across the street purchased a 100-foot hose and invited gardeners to fill water barrels from the shop's taps. The garden provided a space for newcomers to meet longtime neighborhood residents, created a pocket of green space on a busy street, and gave people a way to invest in the community. However, community cohesion, never well established, failed completely. Interpersonal and organizational conflict made it impossible to communicate and reconcile disparate expectations for the garden. Tensions around race and tenure in the neighborhood—clear and unmistakable throughout the Brooklyn Community Garden's establishment and decline—resulted in conflicts of governance that thwarted the ability of the garden to build community among a diverse group of residents.

RACE AND NEIGHBORHOOD TENURE

Racial disputes overlapped with contestations around neighborhood tenure, as most of the long-standing residents were black and most of the newcomers were white. The Brooklyn Community Association had long been a black organization in a black neighborhood, but the new white people were warmly welcomed at the association's monthly meetings. The president of the association, an African American woman in her early eighties, was fond of saying things like "as long as you care about making this a good place to live, you are welcome here." At one meeting, Melissa spoke of how people involved in the garden were "people who walked past, people we had never talked to, people not in the Brooklyn Community Association"—a code for the new white residents who were seen by some long-standing residents as disengaged from neighborhood affairs. Still, length of residence in the neighborhood was often used as a marker of legitimacy. At Brooklyn Community Association meetings, for example, speakers tended to introduce themselves with their name and how many years they had lived in the neighborhood.

Despite formal proclamations of diversity, members of the garden used racialized and tenure-based language to express frustration and explain much of the tension. For instance, Joshua's early equation of hating dogs with hating black people saw him trying to expand racism into an area (animal rights) where it lost all meaning. Joshua's appropriation of racism—telling Melissa she was "bigoted" against dogs—was received as extremely insensitive and racist. This incident reverberated through future garden interactions, as in the meeting where Howard brought it up, calling Joshua a racist when disagreeing with his opinion. Issues of tenure led to conflicts over how space should be organized and what activities should take place, such as Cliff's insistence that he be allowed to plant directly into the ground while claiming that he had always grown vegetables in Brooklyn soil with no health issues, while other gardeners wanted to follow up-to-date urban gardening convention and build raised beds. Tish's conflicts with other gardeners were mediated through her identification with being a longtime resident. She expressed relief that she had not been "Cliffed" and had her bulbs uprooted, and she reacted harshly to concerns about noise, writing, "I grew up here! Don't tell me about the Avenue!" Howard, too, used duration of neighborhood presence to indicate legitimacy when he said that a separation of the garden from the community association would be "erasing history."

Though racial diversity and contact between new and established residents are often presented as unqualified goods, not everyone approaches these interactions from the same position of power. Whereas many of the white newcomers to the neighborhood saw their participation in the garden as a way to build connections with long-standing residents, some of the African American and Caribbean people who had lived in the area for decades saw the arrival of this group as a threat to neighborhood stability. The garden was celebrated as a space and project where neighborhood residents of different races could work alongside each other, but the act of gardening together did not instantly produce a shared community.

CONFLICTS OF GOVERNANCE

While it was possible for gardeners to downplay or ignore racial and neighborhood tenure tensions and still participate in the garden, the conflicts over governance that came out of these tensions made the garden a fraught space. Many of the conflicts of ownership were intertwined with both the racial tensions and issues of neighborhood tenure, as they constructed a divide between Melissa—an African American woman who had long been a resident of the area and active member of the Brooklyn Community Association—and her supporters, and the predominantly white newcomers who saw their participation in the garden as conferring a right to shape its direction. Susan Ruddick (1996) writes about the "interlocking systems" of race, class, and gender, noting the way that different identities intersect with each other and with geographic space. Here, we see race interlocking with neighborhood tenure, creating categories of people with different understandings of ownership and leadership of a neighborhood resource.

Without consistent and agreed-upon leadership, the garden lingered in anarchic nongovernance. Many garden members did not want to be involved in the disputes; they tended to their plants and enjoyed the space but did not participate in the e-mail list or meetings. These were the people who bristled when the membership committee presented their guidelines—they did not feel like others had the legitimacy to tell them what to do. Despite their desire to remain outside the conflicts and tension, this group's lack of participation made it impossible to form the type of structure for garden governance that is necessary to address and resolve tensions.

Ultimately, these overlapping conflicts weakened the potential for the garden to create and enhance a shared sense of community. Community gardens do not necessarily facilitate community goodwill and cooperation. This single garden in a gentrifying Brooklyn neighborhood shows that the opposite can indeed occur: rather than being a place where diverse residents came together and diffused racial tension, the Brooklyn Community Garden actually focused those tensions. Effective organization and communication are difficult in all community settings, but the rise and fall of the Brooklyn Community Garden is not just a story of individual personality clashes. Here, participants' sensitivity to race and tenure differences among neighborhood residents framed the conflicts over governance, leadership, and participation.

Proximity is not community; true community takes time to develop. Overcoming racial and tenure-related divisions requires relationships of trust and respect that can be built only through intentional effort devoted to identifying issues, nurturing leadership, and working through conflicts. These relationships were not present in the Brooklyn Community Garden at the time of its destruction. Issues related to leadership, organization, and ostensibly simple garden business (plot allocation, membership responsibilities) became insurmountable as they became bound up with lingering issues of race and duration of neighborhood residence.

FINAL LESSONS

Though much of the community gardening literature is full of successful projects, this story offers a counterexample. Community gardens do not inherently build cohesive communities; to do so requires strong and deliberate governance, active participation, and a willingness to address difficult issues, and these are possible only in situations of mutual trust and support. Whether a garden's tensions revolve around race, class, gender, land ownership, noise sensitivity, aesthetics, or politics, the garden will not turn them into flowers or tomatoes, despite the common refrains that community gardens do just that.

Here, I offer two specific lessons for those interested in building a garden. First, organizers must not underestimate the importance of a defined governance structure, with clear guidelines around membership, leadership, and decision making. Whatever structure is chosen (be it consensus based, democratic, top down, or something else entirely), what matters is that there is a structure at all (Freeman 1972), even though establishing governance and leadership is likely to engender initial conflict. Further, gardens can be spaces where there are either too many rules or too few; finding the appropriate balance is not easy, but it is imperative. This requires setting aside enough time in the winter—before the commotion of the growing season—to consider various structures and their implications. Looking to other gardens for sample rules, systems, and guidelines can be a helpful place to start.

Second, do not undertake the community garden project without carefully considering how preexisting stresses might be exacerbated by the garden. Payne and Fryman's guide to community building through community gardening (2001) stresses that "patient and sometimes difficult community organizing must precede and follow any miraculous overnight transformations" (5). This process allows for an identification and understanding of all the assets and obstacles to a garden's flourishing. As the Brooklyn Community Garden shows, a garden can reflect an already-fractured community; gardening will not necessarily heal it. It is essential to initiate a process of interaction and communication in order to build the trust among participants that is required for negotiating roles and responsibilities.

The Brooklyn Community Garden is likely not a unique case. This raises the question of whether we should be wary of all the warm, fuzzy stories of community gardens successfully uniting diverse groups and building neighborhood strength and cohesion. There is a great deal of advocacy for community gardening, as supporters hail its ability to provide produce, rehabilitate neglected space, create intercultural bonds, and act as a focal point for neighborhood organizing—one particular *Wall Street Journal* article about a community garden established in the South Bronx explicitly mentions all of these benefits (Huntsberry 2014)—but we must ask why stories of conflict, tension, failure, and disappointment are so rarely mentioned. Is it because gardens are under constant threat of being bulldozed for development (see, for instance, Kennedy 2009 and Alberts 2014) that they must be presented only as an unqualified good? Guthman (2008) writes that many scholars "want to support [the] nascent 'food justice' movement and are cautious about characterizing it in any but the most adulatory ways" (432); it is possible that by doing this we fail to recognize—and fail to learn from—the difficult parts of growing community in a garden.

NOTES

1. Not the garden's real name.
2. In 2000, African Americans made up 78 percent of the population in the community district; by 2010 this was down to 65.5 percent. In the same period, the white population of the neighborhood increased from 6.8 percent to 16.7 percent. In the four census tracts surrounding the garden, the African American population dropped from over 13,000 to just under 10,000, while the white population increased from about 750 to over 4,000.
3. Discussions of the neighborhood's change happened on a Brooklyn-wide community message board with sections for various neighborhoods, neighborhood-specific Facebook groups, news sites including the Huffington Post and DNAInfo, and the Brooklyn real estate site Brownstoner.com.

4. A pseudonym.
5. This and all other names are pseudonyms.
6. Because I attended most garden meetings and many committee meetings, I was perceived as being very involved in the garden.
7. Volunteering is a fairly standard requirement for gaining priority for plots at popular community gardens. See, for example, the rules from another Brooklyn community garden, which state that volunteering is a requirement for membership, and only members are eligible for private plots in which to garden (Prospect Heights Community Farm 2014).

DISCUSSION QUESTIONS

1. How can/should the city itself facilitate urban gardening structures?
2. What sort of role did the Internet play in the daily and overall activities of the Brooklyn Community Garden? Is the Internet a vital resource for community gardens? Why or why not?
3. The "anarchic" initial structure of the Brooklyn Community Garden is atypical of most community garden governance structures, and Thrasher credits this structure with contributing to the diversity of the garden. What role does governance structure play in determining who participates?
4. The incident between Melissa and Joshua occurred while discussing the formation of a potential rules committee for the Brooklyn Community Garden. What greater challenges for the garden did this particular spat indicate? What can gardens do to simultaneously foster diversity while also minimizing racial tension?
5. How could the Brooklyn Community Garden develop a more organized governance structure while also keeping the benefits of a diverse and expansive set of participants and members?
6. What could/should the Brooklyn Community Association and the Brooklyn Community Garden have done with their relationship to improve the welfare of the garden? Should they have had a stronger relationship from the get-go? No relationship? Why?
7. The tension involving the unintentional slight to Cliff was never

resolved. What could be done to address intergenerational conflict in the future?
8. If the Brooklyn Community Garden had successfully established a highly structured governance structure, would it be around today? Would it be at all the same organization?
9. Were racial and neighborhood insider/outsider tensions a result of miscommunications or intended antagonism from any involved party?

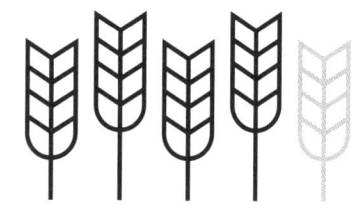

SECTION FOUR
Distribution

While much attention has been given to issues of urban production, until recently far less effort has been put toward understanding distribution systems that work for local foods in urban areas. If production is for hyperlocal consumption or direct marketing only, then distribution is perhaps not as much of a concern; however, time and energy spent distributing produce in urban areas can quickly become prohibitive for farms that market in several different locations or retail outlets. Peri-urban farms with connections to urban markets may find distribution to be the primary hurdle to increasing sales, and urban farms spread out across multiple lots in a city may find transportation delays and expenses a significant barrier to expanding. While most urban-agriculture practitioners do not seek to scale up to the level of conventional wholesale markets, many are seeking ways to streamline distribution, learning from mainstream distributors while preserving the values of alternative food movements.

The case study of the Regional Access company in chapter 11 demonstrates the synergies created when urban food systems are linked to peri-urban and more rural farms within the region. This acknowledges that cities will not produce enough food to meet their food needs but provides an alternative to the conventional supply chain. Regional Access provides an example of a company that provides aggregation, warehousing, distribution, and marketing infrastructure—the same services that conventional supply chains offer—while maintaining the farm-to-consumer relationships and trust that are so key to local food systems. Chapter 12 describes a marketing innovation that has increased food access in areas conventional supermarkets have abandoned. Rather than trying to attract only supermarkets to these neighborhoods, the city of Chicago has supported farmers markets, returning, in a way, to the historical reliance on public markets to deliver fresh foods to cities. The recent prominence of the supermarket has masked the many other possible solutions to food access, including farmers markets, corner stores, and other small retail outlets, which may be able to address food insecurity more rapidly and more appropriately than larger grocery chains. Chapter 12 provides a framework for assessing issues of food access that is more complete than simple distance to grocery stores, and an example of how this can be applied in a city like Chicago.

CHAPTER 11

Food Hubs

Expanding Local Food to Urban Consumers

BECCA B. R. JABLONSKI AND TODD M. SCHMIT

INTRODUCTION

Despite a renaissance of urban agriculture in the United States as described by other authors throughout this book, the overall volume and value of food currently produced in these spaces is unclear. As Johnson, Aussenberg, and Cowan (2013) remark, "there is no compiled [U.S. Department of Agriculture (USDA)] . . . data specific to farms located in urbanized areas" (12). Rather, USDA data are based on "metropolitan" (metro) areas, which include a larger area than just "urban." Though USDA data from 2008 show that more than one-half of all farms located in metro counties have local food sales, compared with only one-third of all US farms, the extent to which these farms are located in an urbanized environment is unknown (Low and Vogel 2011; Johnson, Aussenberg, and Cowan 2013). Further, despite a growing number of case studies documenting the proliferation of or potential for urban community gardens and farms (e.g., Ackerman 2011; Hodgson, Campbell, and Bailkey 2011; PolicyLink 2013), collectively these reports convey that only a small percentage of metro farm production output occurs within urban centers.

As one might expect, available data demonstrate that local food market outlets and sales are concentrated in urban areas (e.g., Hinrichs and Charles 2012; Jablonski 2014; Low and Vogel 2011; USDA ERS 2012). Additionally, Low and Vogel (2011) show that the majority of local food is sold through intermediated markets (defined by King et al. 2010 as a supply chain for a local product that reaches consumers through one or more intermediaries). This suggests that continued growth in local food sales will require intermediaries to move product from farm to market. The widespread agreement that there remains unmet demand for locally grown food (e.g., Hardesty 2008; Baker, Hamshaw, and Kolodinsky 2009; Stephenson and Lev 2004; Schneider and Francis 2005) implies that there is a failure at some point (or multiple points) along the supply chain. Based on a comprehensive literature review, Martinez et al. (2010) conclude that the unmet demand is largely a result of the "lack of distribution systems for moving local foods into mainstream markets" (iv).

Accessing appropriately scaled markets is difficult for small and midsized farms as supply chains become more vertically integrated and consolidated. Large-scale supermarket retail and wholesale operations demand large volumes, low prices, and consistent quantities and qualities that must meet increasingly strict safety standards. The procurement systems in such markets are often vertically and horizontally integrated and global in scale, and they aim to maximize efficiency (e.g., King et al. 2010; Richards and Pofahl 2010; Sexton 2010; Tropp, Ragland, and Barham 2008).

In order to facilitate market access for small and midscale farms and improve consumer access to locally grown foods, public agencies and private foundations are increasingly financing and promoting "food hub" development (e.g., NGFN 2013; USDA 2011; Cuomo 2013). Following the USDA's working definition, a food hub is a "business or organization that actively manages the aggregation, distribution, and marketing of source-identified food products primarily from local and regional producers to strengthen their ability to satisfy wholesale, retail, and institutional demand" (Barham et al. 2012, 4).

Despite the increase in public and private support for food hubs, there has been little work to evaluate their impact. Efforts to assess the impacts of local food system activities generally, and food hubs specifically, are often complicated by a lack of available data. The primary objective of this chapter is to better understand the extent to which food hubs increase consumer access to locally grown and processed products, enhance farm entry into markets, and support farm viability. Given the significant data needs to conduct this type of analysis, we used a case study approach, examining a food hub in New York State (NYS). Accordingly, we conducted surveys with over 300 of the food hub's customers, and in-depth interviews with 30 farms and 15 processors supplying food products to the food hub.

The rest of this chapter begins with a review of the literature related to assessment of food hub impacts on overall supply of local food and farm viability. Next, we present information about our methodology and case study food hub. We conclude with our results, discussion, and recommendations for future research.

LITERATURE REVIEW

There is a burgeoning literature on food hubs, focused mostly on descriptive statistics (e.g., Barham et al. 2012; Bloom and Hinrichs 2011; Clancy and Ruhf 2010; Matteson, Gerencer, and Pirro 2013; Fischer et al. 2013) or their feasibility, best practices, and opportunities to support farm and community development (e.g., Abatekassa and Peterson 2011; Conner et al. 2008; Diamond and Barham 2011; Feenstra et al. 2011; Hardesty et al. 2014; Hoshide 2007; Jarosz 2000; King and Venturini 2005; Stevenson and Pirog 2008; Day Farnsworth and Morales 2011). The espoused benefits of food hubs include the following: (1) expanded farm access to markets, particularly for midscale producers; (2) price premiums for farms through maintained source identification; (3) decreased market costs for farms through resource sharing; and (4) better access to healthy fruits and vegetables for low-income consumers (e.g., Cohen and Derryck 2011; Diamond and Barham 2011; Hoshide 2007; Jablonski, Perez-Burgos, and Gómez 2011; King et al. 2010; Painter 2007; Stevenson and Pirog 2008; Trauger 2009; Schmidt et al. 2011).

Though this literature is useful in many respects, as a whole it does not provide a critical perspective with enough information to discern the ability of food hubs to expand total local food availability or assess the impact of food hub development on farm profitability. Pointedly, Boys and Hughes

(2013) find that scant research exists that explores the extent to which local food infrastructure development "cannibalizes" sales from other markets. They find that the limited available evidence points to what Thilmany et al. (2005) refer to as a "beggar thy neighbor" phenomenon taking place, where support for new market development results in diverted sales from other, previously established markets.

The extent to which food hub development displaces other local food sales has not been well explored. King et al. (2010) studied 15 supply chains in 5 US states and show that locally grown food is available from mainstream, intermediate, and direct markets. Similarly, Abatekassa and Peterson (2011) identify competition between local independent food retailers and alternative market outlets for local food producers in southeast Michigan. Neither study, however, analyzes the extent to which growth in sales in one market impacts sales in another, or how the competition between markets affects total local food availability or farm viability.

Several studies discuss farm-level impacts resulting from food hub development, although they are limited in their quantitative assessments. Schmidt et al. (2011), for example, found many farm-level benefits in their study of a Vermont-based food hub, including price premiums over other wholesale channels. While they note that "many farmers reported an increase in their farm's food production, sales and income" because of their involvement with the food hub, they do not elaborate on this point (e.g., how many farms? by how much?). Furthermore, the benefits are less clear given that the hub, at that time, had not yet reached a break-even point for operations and was supported by public and private grants. Jablonski, Perez-Burgos, and Gómez (2011) suggest in their study of an upstate New York food hub that hubs can reduce marketing and distribution costs for farmers, showing explicit results for an existing farm participating in the hub. However, the narrow scope of analysis limits the generalizations of their results, and again, the food hub was grant dependent.

King et al. (2010) provide the only strong evidence that localized supply chains can enhance farm viability. They find that farms receive a greater share of retail prices in local food supply chains than in mainstream chains, with "net revenue per unit in local chains rang[ing] from about equal to more than seven times the price received in mainstream chains" (v). However, their report does not delineate the extent to which the availability of localized supply chains expands overall farm sales (versus diverting sales from one market to another) or consumer access to locally grown goods.

METHODOLOGY: CASE STUDY

We follow a case study approach, given the detailed data needs to conduct this type of assessment. We chose Regional Access, LLC (RA), because it fits within the USDA's regional food hub definition as an aggregation and distribution business that is committed to supporting local farmers and preserving source identification. In addition, RA's length of time in operation, diverse customer base, and operational size make it a useful food hub to examine. Established in 1989, RA had over $6 million in sales in 2011 and employed 32 full-time equivalent employees. With 9 vehicles and a 25,000-square-foot warehouse, RA aggregates and delivers products primarily throughout NYS, with over 3,400 product listings, including beverages, breads, cereals, flour, meats, produce, prepared foods, grains, and fruits and vegetables.

RA purchases products directly from 96 farm vendors and 65 specialty processors (nonfarm vendors), as well as larger-scale food service distributors. The products are sold to over 600 customers, including individual households, restaurants, institutions, other distributors, fraternities and sororities, buying clubs, retailers, manufacturers, and bakeries. RA also provides freight services to a range of businesses. We conducted in-person interviews with 30 of RA's farm vendors (out of 86 in NYS), as well as with 15 nonfarm vendors (out of 55 in NYS).

It is very difficult to discern the precise role that a food hub plays in terms of enhanced farm viability or profitability, as each aspect of the supply chain may play an important role (e.g., without access to credit a farm is unlikely to be viable even if it has terrific market access). Given this challenge, we determined to understand the farmer's perspective of the role that the food hub played in supporting business expansion or viability. Vendors were therefore asked a series of questions related to the following: (1) the percentage of their total sales facilitated by RA; (2) the extent to which their relationship with RA expanded their overall business; and (3) other key pieces of infrastructure that facilitated farm business expansion or viability.

The 30 farms interviewed were located in every region of NYS except New York City and Long Island. Of the farms from which RA purchased product, 50% classified their operation as "small" ($1,000–$249,999 in gross sales), 20% as "medium" ($250,000–$500,000 in gross sales), and 30% as "large" (over $500,000 in gross sales). When asked to classify their farms' primary production category, 37% identified meat and livestock, 30% fruits and vegetables, and 33% value added (e.g., honey, maple syrup, salsa made from tomatoes grown on the farm).

The 15 nonfarm vendors classified their primary business function as animal slaughtering and processing (3), frozen food manufacturing (3), snack food manufacturing (2), perishable prepared food manufacturing (2), grain and oilseed milling (1), soap and lotion manufacturing (1), wine wholesaling (1), bakery (1), and canning and pickling (1). On average, nonfarm vendors sold over $4.2 million in products in 2011 (ranging from $4,300 to $42 million) and had 32 full-time equivalent employees (ranging from 0 to 320).

RA's customers were surveyed using a combination of an online survey and follow-up phone interviews. At the time of the survey, RA customers included 110 households and 547 businesses, of which 57 and 248, respectively, responded to the survey (a 46% aggregate response rate). Customers were asked a series of questions related to the following: (1) the expanded availability of local or source-identified products as a result of their relationship with RA, (2) the reduction in purchases from other sources as a result of their relationship with RA, (3) the potential for expanded local or source-identified purchases from RA (or another local food distribution company), (4) price premiums customers were willing to pay for local items (business customers only), and (5) unmet demand for local items.

Average annual gross sales for business customers was $5.7 million (N = 101), with a range from $3,000 to $414 million. On average, they had been in business 13 years, ranging from new to more than 130 years (N = 151). The average number of full-time equivalent employees was 15 (N = 145). According to primary business function, business customers identified themselves as distributors (2%), grocery/meal delivery service providers (3%), processors/manufacturers (9%), wholesalers (11%), restaurants (25%), retailers (34%), and other (17%) (N = 245).

RESULTS

Our results are divided into 3 sections—the first 2 examine how the food hub impacted the availability of local food, looking specifically at the impact on consumer access and at the farm and nonfarm vendor access to markets. The third section examines farmers' perception of the impact that food hub sales had on farm viability.

Impact on the availability of local food: Consumer access

Both household and business consumers reported increased availability of locally grown and processed items because of the existence of RA. The majority of customers (62% of households, 57% of businesses) responded that if RA did not exist, they would not know or would be unsure of a place to purchase similar products. Almost 80% of businesses reported that working with RA enabled them to expand their product offerings, on average by 31%.

Both household and business customers reported additional demand for the types of items carried by RA. Though over one-half of household customers (51%) said they were unsure or did not know whether they would purchase additional items from RA if it expanded, 33% said they would. Business customers were more interested (67%) in making additional purchases if RA expanded its product availability, delivery routes, or times.

Businesses were also asked how the price they received for items marked "locally grown" compared to the price for similar, nonlocal items. Based on a scale from 1 to 5 (1 = significantly lower price for local, 3 = no price difference, 5 = significantly higher price for local), on average, customers responded that they received a modest price premium for items labeled locally grown (3.49); overall, 3% reported significantly higher prices, 49% somewhat higher prices, 42% no price difference, 5% somewhat lower prices, and 1% significantly lower prices.

Impact on the availability of local food:
Farm and nonfarm vendor access to markets

Farm and nonfarm vendors reported enhanced market access because of their relationship with RA. Of the farm vendors, those that were midscale (farms with gross sales between $250,000 and $500,000) reported being most reliant on RA's services (fig. 14). All 6 of the midscale farmers interviewed reported that over 20% of their farm's total sales were facilitated by RA—3 had 20%–50%, and 3 had over 50%. Of the small farms interviewed (those that earned under $250,000 in gross sales), 6 had less than 20% of their total gross sales facilitated by RA, 3 had 20%–50%, and 6 had over 50%. Large farms (those with over $500,000 in gross annual sales) reported less reliance on RA-facilitated sales overall, but with a relatively bimodal distribution. Six reported that less than 5% of total sales were facilitated by RA, and 1 less than 20%, while the remaining 2 firms reported facilitated sales of 51% and 93%.

Grouping facilitated sales by primary farm commodity—that is, livestock, fruits and vegetables, and value added (fig. 14)—shows a somewhat more even distribution. Between 27% and 56% of producers had less than 20% of their sales facilitated by RA (fruits and vegetables 56%, value added 50%, livestock 27%), while 33% to 40% had over 50% of their sales facilitated by RA. Livestock producers had a larger share of the midsales range (20%–50%) facilitated by RA, compared to fruit and vegetable or value-added producers (36%, 11%, and 10%, respectively).

Turning to nonfarm vendors, smaller businesses, particularly those with less than $500,000 in gross sales, were more reliant on RA (fig. 14). Of those, over one-half had at least 50% of their sales facilitated by RA. None of the nonfarm vendors with over $500,000 in gross annual sales had over 50% of their sales facilitated by RA, and only 20% had between 20% and 50% of their sales facilitated by RA. The rest of these large nonfarm vendors (80%) reported less than 20% of their sales facilitated by RA.

Impact on farm viability

Of the farm vendors interviewed, 60% reported that their business relationship with RA enabled their business to expand, while 10% were unsure. Of the 30% of farm vendors who responded that RA had not enabled their business to expand, 2 reported that their businesses were not interested in expanding, and another 5 mentioned the importance of gaining access to the New York City (NYC) market through RA, presumably to reallocate sales to better alternatives. Only 1 farm vendor with over $1,000,000 in gross sales

(a) Percentage of farm vendors by size category

(b) Percentage of farm vendors by commodity category

(c) Percentage of nonfarm vendors by size category

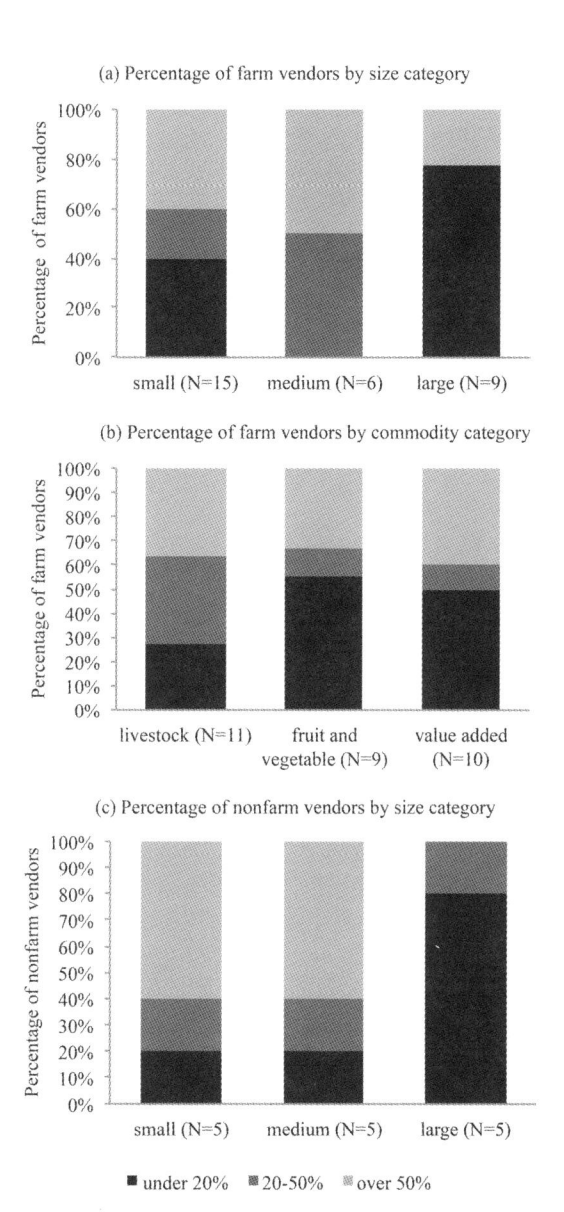

■ under 20% ■ 20-50% ▨ over 50%

Figure 14. Percentage of vendors by level of facilitated sales with Regional Access: A, farm vendors by size category (small = less than $250,000 in sales, medium = $250,000 to $500,000, and large = more than $500,000); B, farm vendors by commodity category (livestock, fruit and vegetable, and value added); C, nonfarm vendors by size category (small = less than $50,000 in sales, medium = $50,000 to $500,000, and large = more than $500,000).

responded affirmatively that RA had enabled his business to expand. Larger farms generally felt that the volume of sales facilitated by RA was too small to make a significant difference in their business's total sales or production, and that they had other market options.

Access to the NYC market was the most frequently cited reason for expanded sales, though improved market access generally was consistently reported. Even farms that were unsure about RA's role in their expanded sales frequently cited RA's freight service and its pick-up and delivery flexibility as the primary reasons farmers chose RA over other freight services to NYC. Others used RA's "good reputation" as a "values-based distributor" to gain market access. This sentiment was particularly true among newer businesses that had not developed direct wholesale purchasing agreements with stores or restaurants.

RA's warehouse capacity was also cited as facilitating business expansion for farms too small to have significant cooler or storage space. Many farms keep frozen meat or storage crops (e.g., potatoes, root vegetables) at RA's warehouse, retrieving them periodically to sell through winter markets, community-supported agriculture (CSA), or wholesale outlets. As a result of access to additional storage, some farmers reported putting more acres into operation, specifically for storage crops, as a way to increase winter (year-round) income.

DISCUSSION AND DIRECTIONS FOR FUTURE RESEARCH

The purpose of this chapter is to better understand the extent to which food hubs increase consumer access to locally grown and processed products, enhance farm entrée to markets, and support farm viability. Our survey results demonstrate that RA increased household and business customer access to locally grown and processed items. Further, the survey results reveal unmet demand for these products, and a consumer willingness to pay a modest premium (on average). Midscale farmers and smaller nonfarm vendors appeared to be particularly reliant on market access facilitated by RA. And the majority of farmers reported that their relationship with RA enhanced their farm's viability, particularly through increased access to the NYC market and warehousing.

Despite these positive impacts, understanding a food hub's impact on farm profitability is very difficult. In this case study, we found that there are numerous ways that RA may support farm viability indirectly. It is difficult to parse out the relative importance of aggregation, warehousing, distribution, and marketing infrastructure provided by a food hub compared to the host of other pieces of infrastructure that farms mentioned were also critical in facilitating sales. Accordingly, future research should make a closer inspection of farm-level impacts from food hub participation. For example, using a market channel assessment approach (see LeRoux et al. 2010 and Hardesty and Leff 2010) to better understand how food hub participation impacts a farm's overall labor requirements, lifestyle, and profitability is critical to evaluating the success of food hubs in farm profitability. Another key area for future research is in considering the opportunity cost—would the farm have found other ways to distribute the product, for a proportionate amount of time and money, had the food hub not existed?

Importantly, these results reflect only one case study. It is possible, for example, that because of the physical location of RA, proximate to the NYC market, there is a particular need for a food hub–type business to expand the availability of locally grown and processed food. More study is needed with additional hubs and locations to assess the robustness of food hub impacts.

ACKNOWLEDGMENTS

This work was supported by Cooperative Agreement Number 12-25-A-5568 with the Agricultural Marketing Service of the US Department of Agriculture; Competitive Grant No. 2012-67011-19957 with the National Institute for Food and Agriculture, USDA; and Grant No. GNE11-021 with the Northeast Region Sustainable Agriculture Research and Education Program. The authors would like to thank Regional Access in assisting in the study design and for access to the financial data and customer and vendor information necessary to complete the analysis. Additionally, we recognize Cornell University Department of City and Regional Planning graduate students Molly Riordan, David West, and Dan Moran for their help with the case study vendor interviews and customer surveys.

DISCUSSION QUESTIONS

1. Jablonski and Schmit found that access to adjacent urban markets was important for rural farmers. Does this finding hold true on a

larger scale; that is, are urban-based local food system initiatives important to farm viability and rural economies?
2. Some survey results from this case study suggested that food retail companies saw potential in selling local produce, as customers were more willing to pay a higher price for it. How could urban agriculturists capitalize on this fact? Should they capitalize on it?
3. Why might medium-sized farms be more reliant on RA than small-sized farms? Or large-sized farms?

CHAPTER 12

Chicago Marketplaces:
Advancing Access to Healthy Food

ANNE ROUBAL AND ALFONSO MORALES

INTRODUCTION

On April 17, 2012, an article by Gina Kolata titled "Food Deserts and Obesity Role Challenged" was published on the front page of the *New York Times*. This article highlighted two very recent studies from the Public Policy Institute of California and the *American Journal of Preventive Medicine* that, respectively, claimed that poor neighborhoods had "nearly twice as many supermarkets and large-scale grocers per square mile" as wealthier neighborhoods, and that there was "no relationship between what type of food students said they ate, what they weighed, and the type of food within a mile and a half of their homes." The *New York Times* article used this research as a foundation for a thesis that essentially dismissed retail food access as an important component of structural health inequity. The overwhelming body of scholarly work suggests otherwise, and the fact that studies can be devised that demonstrate seeming contradictions illustrates the need for this chapter's goal, which is to understand the need for the development of a better metric and the actual conditions and correlates of urban food access (Zenk et al. 2005; Evans 2004; Eisenhauer 2001).

Food access is an often-cited yet ill-defined concept in urban studies, and attempting to define it is surprisingly fraught with complications. Chief among these obstacles is the use of the word "access," which by its nature is an *individualized* notion, yet when used with "food" is often taken to imply a *community* (or even larger) scale. Past definitions and subsequent measures of food access have cited type and scale of purchasing location and distance to the purchasing location as essential components of a definition, with most focusing on supermarkets and grocery stores as primary points of food access. But all food purchasing locations are part of the larger picture of food access, and spatial measurements are subject to their own local meanings based on the individual conditions of the community. When we speak about "food insecurity" and "lack of access," we discuss the problem on a neighborhood, household, and individual scale. Even with the recognition that food access is a massive, national-scale problem, one cannot ignore the importance of the local scale to an understanding of how access functions.

In the chapter ahead we critique measures of food access that care simply about storefronts. Next we turn to our Chicago case to show how farmers markets have been transformed from a neighborhood amenity to an organizational tool that enhances food access. Then, we provide an empirical analysis of farmers markets that demonstrates that they can be an important tool for increasing healthy food access. Finally, we discuss implications of this research for food access metrics. If the actual goal of defining and measuring food access is ultimately to improve conditions within struggling communities, then the first step must be ensuring that research and measurement efforts align with a meaningful understanding of access in the field of study.

FOOD ACCESS: NOT JUST ABOUT SUPERMARKETS

In the same way that public health is not simply about access to hospitals, food access is not simply about access to supermarkets. How urban populations access their food has been an important issue for a century or more (Morales 2011), but scholarly concern accelerated in 2002 when the journal *Urban Studies* featured an issue on urban food access. Researchers in Great Britain led by Neil Wrigley used the term "food desert" to characterize neighborhoods in the city of Leeds with low access to supermarkets (Wrigley 2002). Though not the first time "food desert" was used with respect to food access—our contemporary usage is attributed to a British research group in 1995 (Wrigley 2002)—this article made the term common in research and practice.

Our point is that the term "food desert" typically refers only to *supermarkets*. Farmers markets are ignored in this definition. Given that supermarkets dominate food retail, most research looks to these businesses as a proxy for food access, and there are many good reasons to do so. The ability of supermarkets to exploit industrial and global food supply chains makes them the principal food-purchasing destination in the United States. This supply-chain dominance enables them to dominate smaller venues on price, selection, and one-stop-shopping capabilities (Tamis 2009). Additionally, data on supermarket sales are maintained by industry databases, which are summarized by the North American Industry Classification System (NAICS).

Yet supermarkets' dominance does not always correlate with food access and has in fact often eroded food access. Eisenhauer (2001) and Wrigley (2002) indicate how economic decisions made in corporate headquarters impact local food context. Over the last fifty years, major grocery chains have sought locations to accommodate larger stores, more parking spaces, and higher profits in the suburbs (Eisenhauer 2001). As it became more profitable for supermarkets to move out of the cities into suburban neighborhoods, opportunities for the purchase of healthy food followed (Cronon 1991). Eisenhauer (2001) refers to this trend as "supermarket redlining," or the process by which corporations follow high-profit areas. Between 1968 and 1984, Hartford, Connecticut, lost eleven of its thirteen grocery chains, and between 1978 and 1984, Safeway, a prominent grocery store, closed more than six hundred inner-city stores around the country (cited in Eisenhauer 2001). Some tout the return of chains to cities, yet research demonstrates that large supermarkets centralize food purchasing in a region. For instance, when a Walmart Supercenter opened in Chicago's Austin neighborhood, some eighty-six small businesses within a four-mile radius closed over the course of two years (Davis et al. 2009).

Outdoors or public or farmers markets are well known for their capacity to integrate social, economic, and food access functions, although the former two are most frequently documented (Widener, Metcalf, and Bar-Yam 2011; Bader et al. 2010). Such markets are *places*, amenities attractive to neighborhood residents because they contribute to quality of life and sociability (Balkin and

Mier 2001; Fried 2005). Tangires (2003) demonstrates how market design influences community use, and Gerend (2007) suggests that markets can provide uses for underutilized and vacant sites. Markets and street vendors contribute to economic and community development by providing a variety of benefits (Morales, Balkin, and Persky 1995). They are instrumental in incubating new businesses, facilitating the expansion of existing businesses, and promoting income-earning opportunities. Historically, markets served various economic, social, or political objectives (Tangires 2003; Pirenne 1925; Morales 2000). In the United States, city governments created public markets to address a variety of social problems, such as ameliorating unemployment, incorporating new immigrants, and improving access to healthy food (Tangires 2003; Deutsch 1904; Morales 2000).

All the benefits of markets listed previously have been well documented and are fairly easy to measure (e.g., economic benefits). Markets play a role in addressing health, ecological, and environmental concerns (for a specific case study see chapter 7). They reduce vehicle miles traveled, enhance local sustainability, and help ensure food security. As renewed interest in healthy and sustainable food continues, markets are reemerging as an integral part of the food system. Markets, as they have done in the past, provide a place for social interaction, economic opportunities for vendors and purchasers, and a location to purchase fresh produce. However, for our immediate purposes access to nutritious food with a focus on improving health outcomes is the most important facet of farmers markets, even if it is markedly difficult to quantify.

Poor nutrition is a risk factor in four of the six leading causes of death in the United States—heart disease, stroke, diabetes, and cancer (Anderson and Smith 2003; Pollan 2008). Additionally, racial and socioeconomic inequalities exist in terms of access to healthy foods, leading to insufficient nutrition and an increase in food-related diseases among at-risk groups (Zenk et al. 2005). We know that what people eat and how they eat contribute significantly to mortality, morbidity, and health-care costs (Pollan 2008); we also know that eating a well-balanced diet correlates with good health (Institute of Medicine 2001). However, adequate access to fresh foods continues to depend on both income and location of residence. Thus, the siting decisions of grocery store owners may lead to myriad public health problems. Increasing access to and affordability of farmers markets is one way to reduce this trend (see chapter 14 for examples of the many other impacts farmers markets can have on health). Here our focus is on farmers markets and the particular role they can play in food access apart from the role of the supermarket.

METHODOLOGY

The primary criterion that comes to mind for many laypeople and scholars when discussing food access is defining a distance from stores that makes one area accessible and another one not (Freedman and Bell 2009). Clarke, Eyre, and Guy (2002) measured square feet of grocery space relative to a household's location, for instance, and Guy, Clarke, and Eyre (2004) provided a spatiotemporal model of grocery store type and location over a twenty-year time frame. But this definition should also consider the various modalities of transportation to a store, including walking. However, the very definition of an accessible walking distance is contentious. In the literature, round-trip distances from five hundred meters (about one-third of a mile) to one mile are referred to as "walkable." Some studies also delineate between access via automotive and nonautomotive forms of transportation (e.g., the Environmental Systems Research Institute [ESRI] metric), where both a ten-minute drive and a one-mile walk are used as buffer distances from stores (Herries 2010). Furthermore, some metrics do not choose to indicate any "ideal" distances, claiming instead that accessible stores and their modes of access should be determined relative to a neighborhood's individual context.

Whatever distance is selected, this is only the start of understanding questions of access. To actually determine real conditions of food access for an area, more nuanced measurements that include less tangible conditions are crucial. These include cultural factors such as cooking knowledge and capabilities, attitudes toward certain retailers or types of food, the cultural appropriateness of food available in an area, and the overall perception of residents regarding their food access. Other important correlates that add to our measure would include race or ethnicity, income distribution, or transit measures (such as car access or location of public transit).

The analysis for this chapter began with attention to less tangible, organizational measures and the policy history of markets. We then examined current market locations with an emphasis on neighborhood demographic factors of income, race, and access to public transit at the census tract level. We used Chicago, Illinois, for our data analysis because of the attention paid to food deserts there. This resulted in a pseudo time series analysis regarding market locations before and after the 2009 increase in public markets in Chicago. At this point the city began to think of markets as important policy tools to improve healthy food access instead of only locations for exchange. This descriptive analysis is important in understanding what barriers exist for food access.

POLICY HISTORY AND EMPIRICAL ANALYSIS

As previously noted, the changing retail climate over the past fifty years has reduced the number of grocery stores and supermarkets within major city centers by processes such as "supermarket redlining." Additionally, individuals who remained in the city were underprivileged minorities and low-income earners. In their multistate study, Morland, Wing, and Diez Roux (2002) found four times as many grocery stores in predominantly white neighborhoods as in predominantly black ones, and King et al. (2004) were among the first to document the higher prices and smaller selection of fresh, whole-grain, and nutritious foods in inner-city supermarkets. The market-driven relocation of grocery stores to the suburbs left behind the conditions for a public health disaster. Chicago is one of a number of cities around the United States combating issues of food insecurity within its boundaries.

A private consulting group produced an influential report regarding food deserts (Gallagher 2006), which spurred policy change in Chicago. The report revealed that nearly 632,974 people in Chicago (23 percent of the population) lived in food deserts. This number included 202,054 children (Gallagher 2006) and was made up nearly exclusively of people in majority African American neighborhoods. The report also highlighted negative health effects associated with living in a food desert. Diabetes prevalence and higher obesity rates were associated with food deserts at the census tract level (Gallagher 2006). As a result of these findings, the city adopted policy changes with the intention of eliminating food deserts by 2020 (Chicago Department of Public Health 2011; Gallagher 2011b). A follow-up report was released in 2011 that indicated

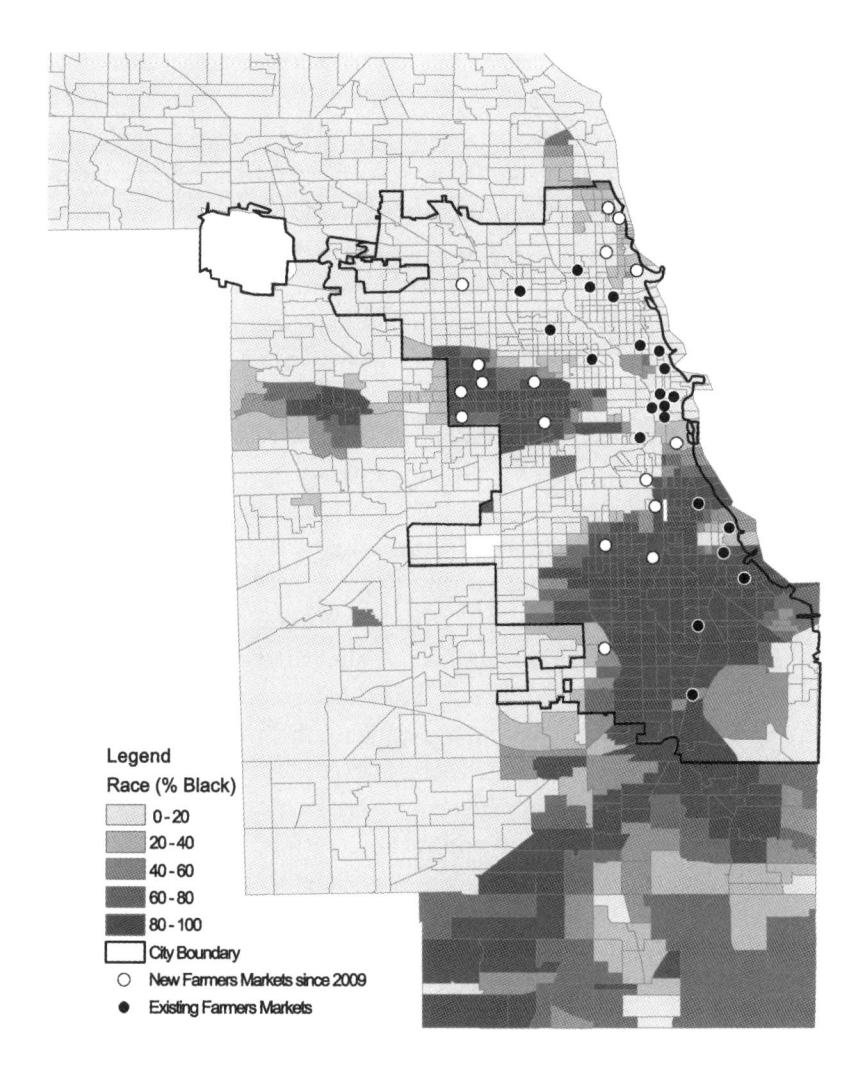

Figure 15. Location of farmers markets in Chicago.

that the number of Chicago residents living in food deserts decreased by 40 percent during the five-year time span (Gallagher 2011b). One of the main reasons for this decrease was the creation of new farmers markets by the city.

We found that farmers markets were most often created in neighborhoods for one of two reasons: one, markets were created and survived in economically viable neighborhoods (in Chicago, these neighborhoods tended to be white, middle- to high-income earners); or two, markets were created after relatively large investments of finances, effort, and time by community-based organizations (CBOs). Markets that have survived in underprivileged neighborhoods, many of which were classified as food deserts, continue to exist only because the operational costs and organization are provided by these CBOs. CBOs are often more prevalent in African American neighborhoods, relative to those of other minorities; thus food deserts are decreasing among this population as the healthy food supply is improved in African American neighborhoods. However, this has left an important sector of the city's population—immigrants in particular, generally of Hispanic/Latino ethnicity—without a proper supply of healthy food. Figure 15 indicates the location of farmers markets before and after 2009, when Chicago introduced its initiative to reduce food insecurity using farmers markets as a tool. Readers will notice the stark contrast between the percentage of African Americans residing in each census block and the formation of the markets.

In order to reduce food insecurity in the city using farmers markets, the city of Chicago must make the markets accessible. Accessibility, as we have alluded to, is often difficult to measure. It undoubtedly must include facets of location, days/hours/months of operation, and availability of healthy foods within the markets themselves. We now use Chicago as an example of how these measures of accessibility may fail to truly identify neighborhoods that have limited access to healthy foods.

As of June 2013, there were forty-four farmers markets operating in Chicago. Of these, twenty-one are city managed, while the rest are independent entities. At least one market in Chicago is operating on every day of the week, although the highest number, thirteen, are open on Saturdays. Markets are generally open during the morning and early afternoon hours with the exception of a few, such as the LaFollette Park Farmers Market, which is open from 1:00 p.m. until 7:00 P.M. (LaFollette Park Farmers Market 2012). Four of the forty-four markets are open in the late afternoon and evening, while the rest are open in the morning hours. Most of the markets operate primarily during the summer months, but a few also operate indoor markets during the winter. The number of patrons at markets also varies greatly. The Green City Market boasts 9,000 unique visitors per day, while the Logan Square Market reports between 2,500 and 3,000 visitors each day (Larmer 2012; Green City Market 2012).

The hours and days of operation are important facets of access, as individuals may need to purchase food on their way home from work or on the weekend when they are running errands. Table 10 depicts other ways that the markets in Chicago differ, such as in the number of vendors and the years, days, and hours of operation. In order for markets to be used as a step to reducing food insecurity, it is important that markets operate at times when consumers are able to purchase food.

Another important facet of accessibility is location. Farmers markets in Chicago during the summer of 2013 were fairly well dispersed geographically. More markets were located near the loop and downtown area in the densest part of the city, making them responsible for serving more people (fig. 15). Markets cover 33 of the 61 zip codes and 41 of the 878 different census tracts in Cook County. Although markets are fairly well dispersed throughout the county, only 63 percent are within a half-mile of a Chicago Transit Authority (CTA) rail line, which is a major mode of transportation throughout the city, especially for low-income earners. The CTA is widely used in Chicago because of its accessibility and low cost relative to owning a vehicle. Twenty-seven farmers markets are within a half mile of a rail line, while seventeen are not. This distinction may result in different purchasing patterns and patrons at the markets. Markets on or near a rail line provide easy travel routes over long distances, which allows a greater range of people to attend. Markets not near a CTA rail line stop may attract a smaller and more local customer base. In order to achieve maximum efficiency, markets should be located so that they are accessible to the greatest number of people.

The neighborhoods in which markets currently exist vary with respect to certain demographic factors. Chicago is known to be one of the most racially and economically segregated cities in the United States (Ahmed and Little 2008). Farmers market locations also follow this trend. The mean income of the census tracts where farmers markets reside is somewhere between

Table 10. Characteristics of farmers markets in Chicago

Market	City operated	Community-based organization supported	Years in operation	Day of week	Time	Link accepted	Vendors/day (average)	Patrons/day (average)
61st Street	No		2008–present	Saturday	9:00 AM – 2:00 PM	Yes	20	
Andersonville	No	Yes	2009–present	Wednesday	3:00 PM – 8:00 PM	Yes	30	1,200
Austin Town Center	Yes	No	2012–present	Thursday	1:00 PM – 7:00 PM	Yes		
Beverly	No	No	2009–present	Sunday	7:00 AM – 1:00 PM	Yes		
Bridgeport	Yes	No	2009–present	Saturday	7:00 AM – 1:00 PM	Yes		
Bronzeville	No		2007–present	Saturday	8:00 AM – 1:00 PM	Yes		400
Chicago's Downtown Farmstand	Yes	Yes	2008–2012	Mon-Sat	11:00 AM – 7:00 PM	Yes		
City Farm Market Stand	No	No	unknown	Tuesday-Friday	1:00 PM – 5:30 PM	No		
Columbus Park	Yes	No	2012–present	Tuesday	1:00 PM – 7:00 PM	Yes		
Covenant Bank - North Lawndale	No	Yes	2006–present	Wednesday	8:00 AM – 1:00 PM	Yes	5	50
Daley Plaza	Yes	No	2003–present	Thursday	7:00 AM – 3:00 PM	Yes		
Division Street	Yes	No	2003–present	Saturday	7:00 AM – 1:00 PM	Yes		
Eden Place	No		2009–present	Saturday	8:00 AM – 3:00 PM	Yes		
Edgewater	No		unknown	Saturday	8:00 AM – 1:00 PM	No	14	
Federal Plaza	Yes	No	2003–present	Tuesday	7:00 AM – 3:00 PM	Yes		
Glenwood Sunday Market	No	Yes	2010–present	Sunday	9:00 AM – 3:00 PM	Yes	24	1,500
Green City Market	No	Yes	1998–present*	Wed/Sat	7:00 AM – 1:00 PM	No	59	9,000
Healing Temple Church	Yes	No	2012–present	Sunday	2:00 PM – 6:00 PM	Yes		
Homegrown Bronzeville	No	Yes	unknown	Sunday	9:00 AM – 1:00 PM	Yes		
Hyde Park	Yes	No	2003–present	Thursday	7:00 AM – 1:00 PM	Yes		

		2007–present	2nd/4th Sunday	9:00 AM – 1:00 PM	No	30+	
Independence Park	No	2007–present	2nd/4th Sunday	9:00 AM – 1:00 PM	No	30+	
Jefferson Park Sunday Market	No	2011–present	Sunday	10:00 AM – 2:00 PM	No	7	500
La Follette Park	Yes	2012–present	Wednesday	1:00 PM – 7:00 PM	Yes		
Lincoln Park	Yes	2003–present	Saturday	7:00 AM – 1:00 PM	No		
Lincoln Square	Yes	2003–present	Tuesday	7:00 AM – 1:00 PM	Yes		
Logan Square	No	2005–present*	Sunday	10:00 AM – 3:00 PM	Yes	55	3,000^
Loyola's Farmers Market	No	2011–present	Monday	3:00 PM – 7:00 PM	Yes	10	200
Mount Ebenezer Baptist Church	Yes	2012–present	Saturday	10:00 AM – 4:00 PM	Yes		
Museum of Contemporary Art/Streeterville	Yes	2003–present	Tuesday	7:00 AM – 3:00 PM	No		
Northcenter	Yes	2003–present	Saturday	7:00 AM – 1:00 PM	No		
Pilsen Community Market	No	2008–present	Sunday	9:00 AM – 3:00 PM	Yes		
Portage Park	No	2009–present	Some Sundays	10:00 AM – 2:00 PM	No	25	1,100
Printer's Row	Yes	2003–present	Saturday	7:00 AM – 1:00 PM	No		
Pullman	Yes	2003–present	Wednesday	7:00 AM – 12:00 PM	Yes		
Rowan Tree Garden Society	No	2009–present	Friday	8:00 AM – 3:00 PM	No		
Seaway Bank Farmers Market	No	2005–present	Wednesday	9:00 AM – 2:00 PM	No		
South Shore	Yes	2003–present	Wednesday	7:00 AM – 1:00 PM	Yes		
Southport Market	Yes	2007–present	Saturday	8:00 AM – 2:00 PM	No		
Uptown Market at Weiss	Yes	2003–present	Thursday	7:30 AM – 12:30 PM	No		
West Humboldt Park Farmers Market & Bazaar	No	2010–present	1 Sat/month	10:00 AM – 2:00 PM	No		
Wheeler Mansion Market	No	2012–present	Wednesday	4:00 PM – 8:00 PM	No		
Wicker Park & Bucktown	Yes	2003–present	Sunday	8:00 AM – 2:00 PM	Yes		
Willis Tower Plaza	Yes	2003–present	Thursday	7:00 AM – 3:00 PM	No		
Wood Street Farm Stand	No	2010–present	Wednesday	1:00 PM – 4:00 PM	Yes	1	

*In 1998 operated out of the Chicago Theater
˄ Indoor market began operation in 2009
˜ 14 in winter
˅ 300-500 in winter

$20,000 and $40,000. Only two farmers markets are in census tracts with median household incomes lower than $20,000 (fig. 16). Three farmers markets are in census tracts with median household incomes greater than $70,000. Eighteen markets lie in census tracts with incomes between $20,000 and $40,000 (City of Chicago Census Maps 2013). By contrast, 5.6 percent of Chicago census tracts have a median household income of less than $20,000, while 23.8 percent have a median household income greater than $70,000. Even more attention should be paid to attempting to put farmers markets in even lower-income neighborhoods where it is already more difficult to raise a family. In census tracts where median household income is below $20,000, close to 100 percent of the population is eligible for nutritional assistance programs. These neighborhoods should be the focus of system-level interventions in order to maximize health improvement. (For more information on changing the built environment see chapter 16 in this volume.) Targeting this population with such an intervention, such as the addition of a farmers market, may help improve eating behaviors.

The markets are slightly better dispersed by neighborhood racial demographics, partially thanks to the five new markets added by the city in 2012. In that year, twenty-four of the forty-four markets were located primarily in African American neighborhoods (50 percent or more black) (see fig. 15). Since African Americans consume fewer fruits and vegetables relative to the general US population, it is imperative to remove the barrier of access in order to reduce disparities. Focusing more farmers markets in areas of the city dominated by the African American population along with providing education regarding how to prepare fruits and vegetables and why fruits and vegetables are beneficial will result in positive outcomes for the city. The city needs to target "high risk" groups (those who consume much less than the recommended amount of fruits and vegetables) because this will result in the greatest outcome for the city's initial investment. The African American population is not the only subpopulation with low access to healthy foods. Other markets have been started with the goal of improving the health of university students as well as hospital employees and visitors through increasing healthy food access.

Cost remains a barrier to healthy food access for many families. In an attempt to provide nutritious foods for those who need it most, the state of Illinois

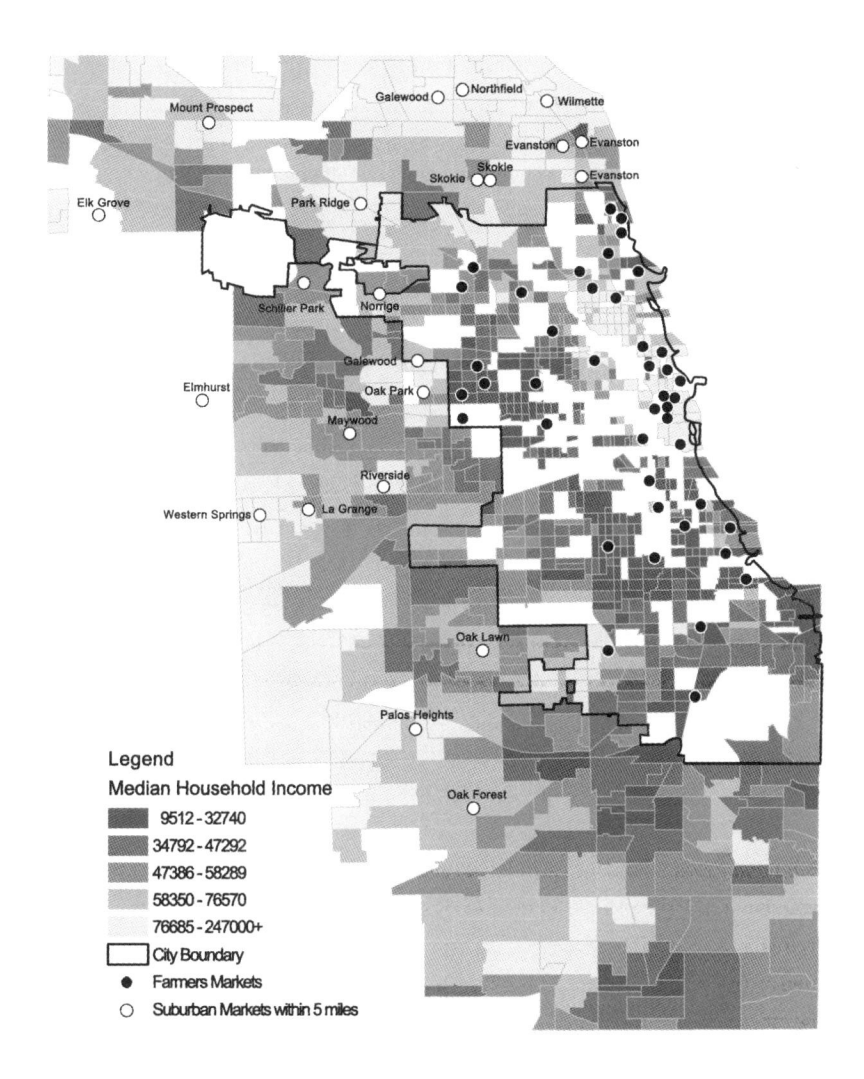

Figure 16. Census tract income.

has created incentive programs for families to maximize their Supplemental Nutrition Assistance Program (SNAP) dollars at farmers markets using their "Link Card." The program is intended to provide money for individuals who cannot afford to purchase food for their family. The benefits come on a debit card with a predefined balance and can be used only for specific food products. Several farmers markets in Chicago have begun accepting Link payment for produce. Additionally, a few markets and government programs have attempted to create other incentives for individuals to spend their Link dollars at farmers markets by doubling the amount of spending money individuals receive if they use it there. In the 2012 season, twenty-seven markets accepted Link as a form of payment, while seventeen did not. Markets such as the one in Andersonville actively promote using Link and provide incentives. The market advertises, "We will match up to $10 in LINK token purchases per card holder, per market day" (Andersonville Farmers Market 2012). The market at 61st Street matches up to twenty-five dollars and provides cooking classes during the market for children (Mission of the 61st Street Market 2012). This serves the purpose of both child care, which assists the consumer, and knowledge diffusion, which is important in encouraging children and parents to eat in healthier ways. The acceptance of Link at markets is important in terms of using farmers markets to reduce food deserts. Many individuals living in food deserts are low income and receive government assistance for food purchases. Making it easier for them to use these funds to purchase healthy foods, such as those sold at farmers markets, will help reduce food deserts and improve health outcomes.

We acknowledge that we have not covered every aspect of food access measures in Chicago. However, developing a better understanding of access measures and the barriers individuals face in accessing healthy and nutritious food will inform the dialogue with the hope that improved measures for food access will be utilized in the future.

MEASURES OF ACCESS

Creating a framework by redefining food access

In order to move toward a more accurate measurement of urban food access, an understanding of the components that create access is necessary. Table 11

Table 11. Comparison of food access measures used in the literature

Metric	Year	Distance	Complex model?*	Database (free)	Database (private)	Other data source
ESRI Food Desert Finder (ESRI); Herries (2010)	2011	1 mile, 10 min. walk	no	no	yes	--
USDA Food Desert Locator (USDA); Ver Ploeg, Nulph, and Williams (2011)	2010	1 mile	no	yes	yes	--
SHAW; Shaw (2006)	2006	--	yes	no	no	survey
Mari Gallagher: Birmingham, AL (MGAL); Gallagher (2010b)	2010	--	no	yes	yes	own
Mari Gallagher: Hamilton County, OH (MGOH); Gallagher (2011a)	2011	--	no	yes	yes	own
Mari Gallagher: Chicago, IL 2011 (MGCHI); Gallagher (2011b)	2011	1 mile	no	yes	yes	own
LEETE; Leete, Bania, and Sparks-Ibanga (2012)	2012	1 km	no	no	yes	--
WID; Widener, Metcalf, and Bar-Yam (2011)	2011	--	yes	yes	yes	--
BADER; Bader et al. (2010)	2010	800 m	yes	no	yes	--
KER; Kershaw et al. (2010)	2010	1 km; 15 min. walk	yes	no	no	--
APP; Apparicio, Cloutier, and Shearmur (2007)	2007	1 km	no	yes	no	--
GUY; Guy, Clarke, and Eyre (2004)	2004	--	no	no	no	--
S-T; Smoyer-Tomic, Spence, and Amrhein (2006)	2006	1 km	no	yes	no	--
CLA; Clarke, Eyre, and Guy (2002)	2002	500 m	no	no	no	survey
HAL; Hallett and McDermott (2011)	2011	500 m	yes	yes	no	--

*"Complex model" here means the presence of other contributing factors in the analysis from distance-based variables.

depicts several of the measures of food access that are used in the literature and provide a comparison across metrics. These metrics are varied, yet each remains useful in its own way. Yet, to advance research a systematic measure that can be used across studies is needed. In an attempt to unify measures Shaw (2006) proposes a three-pronged model for defining food access: ability, assets, and attitude. *Ability* is a function of one's transportation options. Those who lack transportation options are unable to travel to stores as far away or as reliably as the average member of the population. Also described by the *ability* metric are those affected by safety challenges between stores and their home. *Asset* problems involve the lack of some asset that inhibits food access. Some examples of this may be lack of financial assets to pay for the more expensive products in an area, lack of funds for sufficient transit fare, or even a grocer's lack of assets for stocking fresh foods. Finally, the concept of *attitude* attempts to capture any sort of cultural or psychological reason for lowered food access, including a lack of culturally appropriate local options or people's inability or unwillingness to cook for themselves.

While Shaw's tripartite approach to food access captures many of the less tangible components of a holistic definition of the concept, and although no definition will be entirely complete, there is one especially noticeable hole in this one. It places most of the burden of food access on consumers and consequently could be seen to blame them when food access is low (especially in the case of attitude-based access problems). Research and historical experience both demonstrate that urban areas, and especially areas of lower income and higher minority concentration, have been *systematically* excluded from food system development by actors other than themselves, including retailers, policy makers, and planners (Richardson et al. 2012; Alwitt and Donley 1997). For a definition of food access to amplify efforts to fix ongoing problems in addition to defining them, it needs to focus on supply-side access building that involves the civic decision makers and retail interests.

With this in mind, we propose a three-part definitional structure similar to Shaw's, but instead we use a more general framework: the "what," the "where," and the "why" of food access. Each component contains a question and an answer.

What is accessible food? Food that is accessible is nutritious, of good quality, available in variety, fairly priced, and culturally appropriate.

Where is accessible food? Accessible food is found within a reasonable distance of consumers, accessible by many modalities of transportation, appropriately scaled to the community, and safe to go to and from.

Why is food accessible? Food is accessible when people know where their food comes from, how their food was produced, and how to prepare (or find) healthy foods for themselves, and when they have agency in what food they obtain. This facet can be aided by CBOs, neighborhood groups, churches, the government, and other agencies, as each has a unique role in food procurement and distribution.

Although these questions and answers cannot cover every facet of food access, they do provide a broad base for defining the concept in a holistic manner, offering a significant improvement over distance-only definitions.

There are three reasons to prefer this framework to the aforementioned ability/asset/attitude conceptualization, and especially over more common simplistic concepts of grocery store distance. First, it makes mention of less tangible aspects of food access, like food knowledge and individual agency, which are often overlooked in location-based frameworks. It also does not exclude any party from participating in accessible food, nor does it focus too much on or implicate consumers in the problems inherent to the built environment in which they happen to reside. And, perhaps most distinctively, this definition is positive without being idealistic. Instead of trying to define the conditions for *poor* food access across myriad local conditions, it provides an attainable and realistic goal of accessible food that can work in any community context.

TOWARD A LOCALIZED METRIC FOR FOOD ACCESS

In light of this more holistic redefining of the concept of food access, several concrete recommendations can be made about the future of food access measurement. The first is that, clearly, the study of urban food access must be undertaken at a localized level in order to fully and effectively capture the more nuanced aspects of the concept. If analysis occurs at a supralocal level, obtaining the needed level of detail about the food environment will become increasingly inefficient at best and impossible at worst. As mentioned previously, finer scales of analysis also tend to be more meaningful to the average citizen and allow for a picture of access that is much closer to the individual scale. This is not to discredit larger-scale access mapping efforts, like the one undertaken by the USDA in 2010. This work certainly generated more widespread attention to food access and painted a broad picture of the problem as a national concern. However, the trend for metrics to try to generalize their localized findings to broader regions is likely a misguided one if the true goal is improving access for areas in need, and big-picture studies like the USDA's simply cannot be used as the only basis for effective policy interventions.

Following from this preference for localization of analysis, there must concurrently be a push for greater support for the collection of food environment data on this scale. As noted previously, the state of freely available databases is pitiful, with no real information given aside from address in most directory resources. To access even a basic categorization of store locations, researchers must buy into the private data warehousing structure, and even with a paid subscription, the provision of a NAICS code provides no insight into real conditions at the stores themselves. For nonsupermarket food access locations, the data predicament is even worse: listings for small stores are likely to be out of date, if they exist at all, and information on nonmainstream food purchasing locations (like farmers markets) is almost never collected in a centralized manner. When researchers do decide to ground-truth their data or support databases with community-based survey feedback, the results are usually kept proprietary because including the human factor requires extra time and expense.

Particularly with these shortcomings in mind, it is not terribly surprising that some of the best food environment inventories have not been made public. But with the demand for data accessibility and transparency being spurred on by nearly universal access to the Internet, perhaps now is the best time to promote inventory projects like these. Food environment inventories not only provide a more accurate foundation for food access research but could also provide benefits even outside the academic and policy spheres. The resource inventory process is a fairly straightforward one, and local residents should be directly involved, considering that they know their environment the best. This allows many opportunities for community involvement and empowerment, especially for youth (Santilli et al. 2011). Additionally, while a customized data structure will likely serve each community best, inventory frameworks like the Nutrition Environment Measures Study (NEMS) can help standardize data across a variety of applications so they can be better shared and compared (Glanz et al. 2007).

Finally, future studies of food access must move away from oversimplified metrics. The fact remains that food access is a highly complex concept, and thus it must follow that the measure of this concept should be equally complicated by integrating multiple variables and factors. Modeling and measuring social conditions is never a neat practice. Distance-based approaches have consistently overstated their metric power in providing a proxy for these conditions. Instead of shying away from messy results in favor of an eye-catching map, researchers need to embrace advanced methods and technologies in order to include as many factors as possible in their access analyses. Furthermore, these variables should be sensitive to the locality in question. If the majority of residents in a neighborhood lack personal automobiles, why place an emphasis on driving times? These kinds of evaluations can come only out of a nuanced understanding of the community, which would already be emphasized by the local scale of analysis and increased efforts for food environment data collection.

CONCLUSION

Urban food access is a complex issue, but actors in the urban policy arena cannot claim to be working toward a just city without addressing these inequities in infrastructure. Existing literature and research forms an indispensable foundation for understanding food access in cities, but in order to move from understanding to action, there must be a shift in the way that measurements are conceptualized and carried out. Metrics that truly measure access must be based in localized and ground-truthed understandings of the food environment, and they must address inherent complexities in their study environments. The work of the USDA to create a food desert map, with uniform measures across all jurisdictions, is commendable. However, it is just the beginning of measuring food access. Local communities must strive to understand the reasons why food insecurity exists and work toward eliminating them, as Chicago did in adding farmers markets in minority neighborhoods. With a working knowledge of what has been done, as well as a more holistic redefinition of the concept of access itself and a reinvented framework for measurement, the field of food access research can begin to move toward analyses that will better effect change on the ground.

ACKNOWLEDGMENTS

The authors wish to extend sincere gratitude to Riley Balikian, Lauren Suerth, and Claire Boyce, who provided invaluable feedback on this chapter. Additionally, this work would not have been possible without participation from many enthusiastic market operators in the Chicago area.

DISCUSSION QUESTIONS

1. Is the definition of "food access" fundamentally different at the individual, household, neighborhood, state, and national levels? Explain some of these differences.
2. How thoroughly can Shaw's three-pronged model as explained by Roubal and Morales touch on the various issues and aspects of food access and food deserts?
3. Being economically successful and being "local" are two fundamental components of a successful farmers market. How important are both of these factors when considering farmers markets in urban settings?
4. Is the current method of implementing farmers markets in low-income and minority neighborhoods in Chicago sustainable? What could be done to ensure the longevity of these markets?
5. Roubal and Morales describe in great detail the opening and closing times of farmers markets, as well as the locations and why the markets operate the way they do. Does a collaboration of markets like that in Chicago develop organically, or does an outside actor or umbrella organization need to guide the process?
6. The authors comment on the lack of data in analyzing food systems. Given the obstacles this presents, how can governments and communities adapt to address these needs?
7. What data need to be systematically gathered to successfully analyze

the success of farmers markets, the various impacts they have on food insecurity, and the roles they play with respect to food insecurity?
8. Imagine yourself in the role of a high-ranking government official in Chicago. What policy recommendations would you make to best support urban agriculture in the city?

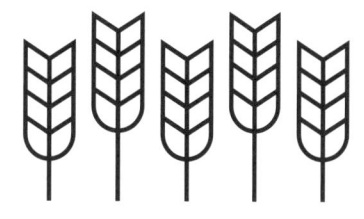

SECTION FIVE

Community Health and Policy Perspectives

It is clear that urban agriculture is about more than just producing food. Urban agriculture is about building community, reconnecting people to their culture and spirituality, improving neighborhood and individual health, and supporting food sovereignty and justice for farmers and communities. These intangible outcomes are very difficult to quantify for planners, funders, and government officials, but this does not make them less real or important.

In chapter 13, Cohen and Wijsman present case studies and strategies for blending practice and policy changes so that policy supports successful practices. In a field as diverse and rapidly evolving as urban agriculture, this dialogue between policy and practice is necessary and highly productive for the future of urban-agriculture practices. No one set of policies will be conducive for all cities, hence the need for a rich literature of case studies, historical examples, and recommendations for best practices in different situations.

Chrisinger and Golden, in chapter 14, review the current state of our knowledge about urban agriculture and public health and then describe possible policy and practices that work through urban agriculture to improve human health. Various practitioners and audiences will find this work useful in considering how to design health interventions, tailor evaluations, or form reasonable expectations about the effects of urban agriculture in their own communities.

In chapter 15, Day Farnsworth provides clear recommendations for organizing food policy councils effectively as well as insights into what types of policies are needed in different contexts. Her work points directly to food production, whether by insects, animals, or the soil. Her point is clear, that food production can be facilitated by a non-food-production organization that locates production in a larger context.

In chapter 16, de la Salle reviews linkages between urban agriculture, the built environment, and population health. Her case studies, review of design approaches, and discussion of best practices provide options for practitioners seeking to improve community health through the built environment and urban agriculture.

The chapters in section five seek to provide guidance for addressing and incorporating these associated benefits of urban agriculture into planning processes. Clearly we need better metrics for outcomes such as improved human health. Through more holistic metrics we can relate health to systems change and the collective impacts of many organizations, institutions, and individuals working on food systems issues. This section of the book provides concepts and examples of activities related to how circumstances, social expectations and pressures, and other institutional spheres, like the legal system and hopes for public health, influence the people doing urban food system activities.

CHAPTER 13

The Coevolution of Urban-Agriculture Practice, Planning, and Policy

NEVIN COHEN AND KATINKA WIJSMAN

INTRODUCTION

After a century of relative inattention to the food system (Pothukuchi and Kaufman 2000), many cities have developed plans and policies to legalize, rationalize, and institutionalize urban agriculture, and to support innovations in food production (Hodgson, Campbell, and Bailkey 2011; Mukherji and Morales 2010; Thibert 2012; Cohen 2012; McClintock 2014). These plans and policies have been influenced by the practices of urban farmers and gardeners, and in a coevolutionary process plans and policies have shaped their practices. Together, practices, plans, and policies support innovative urban-agriculture systems and thus change what is considered the normal, acceptable, and fair way of using space in the city and producing food. As a result, over the past decade food production has become more fully embedded in urban spaces, daily activities, and government programs, including those not traditionally connected to the food system.

This is evident in New York City, which has one of the largest and most diverse urban-agriculture systems in the United States (Cohen, Reynolds, and Sanghvi 2012; Altman et al. 2014). A case study of the coevolution of urban-agriculture practices and plans in New York City illustrates the entanglements of urban-agriculture practitioners, advocates, planners, and policy makers; innovative food production practices that have changed the meaning of urban agriculture; and innovations in practice by city government that have supported urban agriculture. The New York case shows that farmers and gardeners, advocates, and government planners and policy makers can steer systems like urban agriculture by identifying and supporting innovative and sustainable practices, while simultaneously challenging undesirable practices that have become entrenched because of institutionalized rules and norms (Loorbach 2007).

COEVOLUTION OF PRACTICE AND POLICY

Practices are simply the routine things people do to achieve various goals. They consist of three elements: (1) competencies, or the knowledge of how to do things; (2) material items, including technologies and infrastructure; and (3) meanings, or the ideologies, goals, and cultural understandings that make certain practices acceptable and normal (Shove 2003; Shove, Pantzar, and Watson 2012). The practice of growing food in the city, for example, requires horticultural knowledge and other competencies; space, soil, water, and other material resources; and a cultural milieu that makes gardening and farming in the city appropriate, desirable, and normal. Practices may consist of discrete actions, like the act of cultivating vegetables, but are typically part of interconnected bundles of practices (Warde 2005), such as cooking, composting, and selling, that affect each other.

   Practices are ubiquitous and largely habitual activities, and though mundane, they can have significant impacts and influence policies by creating demands on infrastructure, public space, natural resources, and public funds. For example, domestic cooking, refrigeration, and dish washing account for one-quarter of household electricity use in the United States (Canning et al. 2010), while food waste from cooking and eating practices amounts to 14.5 percent of municipal waste streams (US EPA 2014). The practice of urban agriculture, bundled as it is to other urban practices, can therefore have a significant effect on a city's sustainability.

Changes to practices like urban farming and gardening do not occur merely as a result of the aggregate choices of individuals. Rather, they result from changes to the three practice elements made by policy makers, practitioners, advocacy groups, and consumers who support new knowledge and competencies, reconfiguring material elements like infrastructure and increasing the acceptance of alternative practices by reshaping their meanings and helping them become normal. New programs or economic incentives can also "recruit" additional practitioners (Shove and Walker 2010) to engage in practices like farming that diverge from the status quo. These efforts can make what might be considered deviant but sustainable practices (e.g., beekeeping or rooftop farming) normal and everyday.

The actors involved in creating and normalizing innovative practices are diverse. For example, government officials may support community gardening networks practicing specific methods of food production or fund new types of urban farms through municipal programs that provide land, material, and tools to gardeners and farmers. Individuals and advocacy groups may start urban-agriculture projects and engage in innovative practices, like offering beekeeping classes, running community-supported agriculture programs, or developing housing with rooftop farms, which then prompt policy responses. Even individuals change policies through their everyday practices, as they raise chickens, prompting health code changes, or take over vacant lots to farm them, prompting new policies allowing food production on public land.

Deviant practices also influence policy change by providing examples of different opportunities to achieve broad public policy goals (Healey 2012), thus influencing the development of plans and policies. Urban-agriculture plans and policies are not created de novo by government fiat but more commonly respond to the practices of farmers, gardeners, and others in the urban-agriculture system. As new practices enroll practitioners and become normalized, government policies and programs often support and stabilize them, creating lasting transformations to urban-agriculture practice as well as to the many related practices that are bundled to farming and gardening. On the other hand, physical, institutional, and informal structures and practices may restrict new practices from flourishing (Loorbach, Frantzeskaki, and Thissen 2011). For example, growing spaces may be limited, agency rules and practices may be entrenched, and other habits and activities may constrain the ability of new practices to take hold. Dynamics between institutional and cultural structures and norms on the one hand and the power of practitioners on the other hand to shift practices notwithstanding existing systems enable some practices to expand while constraining others.

COEVOLUTIONARY PROCESSES TO SUPPORT URBAN AGRICULTURE

Three examples from New York City's urban-agriculture system illustrate how practices, plans, and policies emerge in interconnected processes. The first shows the entanglements of different actors in the practice of finding and allocating new space for gardens and farms, and the role of practice in changing policies. The second involves innovative practices and supportive policies that have broadened the notion of how and where urban agriculture can be practiced. The third illustrates how practice innovation in both urban agriculture and government can establish a role for agriculture in two key urban systems conventionally unrelated to food—affordable housing and stormwater management—and in so doing connect the practice of growing food across municipal agencies.

COMMUNITY GARDENING PRACTICES AND PUBLIC SPACE

In New York, urban agriculture has grown thanks to the entrepreneurial practices of individual farmers and gardeners, advocacy efforts of gardeners and nonprofit organizations, and supportive public policies and city agency administrative actions. While community gardens and urban farms have always existed in New York City, the number of gardens expanded in the 1970s as the city's fiscal crisis and disinvestment in low-income communities of color resulted in property abandonment and cutbacks in municipal services. Residents of these neighborhoods took over vacant properties, sometimes illegally, cleaned them up, and turned them into community gardens. Throughout this period, the practice of "guerrilla gardening," occupying and gardening abandoned sites, was normalized through the creation of a municipal program in 1978 called Operation GreenThumb, which provided gardeners with material resources, technical assistance, and recognition by the city.

A significant shift happened two decades later when Mayor Giuliani attempted to sell the city-owned community garden sites to developers to build housing (Elder 2005) but was met with political opposition and legal challenges to save the gardens. A lawsuit by the attorney general created an opening for the New York Restoration Project (NYRP) and the Trust for Public Land (TPL) to raise money to purchase some 163 gardens that were to be put up for auction, and in response the city sold them to the two groups and relented on selling most of the others. The outcomes of this battle included galvanizing support for garden preservation by the city's urban-agriculture, environmental, and civic organizations as well as increasing the visibility of the community gardens, the plight of which was covered by the media, which in turn increased the political salience of urban agriculture in New York. The purchase of the gardens by the NYRP and TPL transformed these garden spaces by providing professional management and permanent tenure, thus shifting the practice of gardening from an activist, grassroots practice to a more established and permanent aspect of city life.

The normalization of community gardening led to policy changes to support and expand the city's urban-agriculture infrastructure. Urban gardens and farms, once treated as a temporary use of city parcels slated for development, have been described in recent city strategy documents (a City Council policy platform called FoodWorks and the administration's sustainability strategy, PlaNYC) as important land uses and valuable activities. In 2010, the city adopted rules to provide renewable licenses for community gardens on public land and an extensive public review process that makes evictions less likely (Cohen, Reynolds, and Sanghvi 2012). The City Council enacted Local Law 48 of 2011, requiring the city to create an online public database of vacant city-owned property that includes an assessment of each parcel's suitability for urban agriculture. Subsequent to the bill's enactment, the city administration identified more than one hundred properties as potentially suitable for urban agriculture, while a second city program, Gardens for Healthy Communities, coordinated by the Department of Parks and Recreation and its GreenThumb program, made twenty plots of land available and provided program support for community groups to turn them into community gardens.

These policy changes, coupled with growing interest nationally in the practice of urban agriculture, have in turn led the New York City Community

Garden Coalition, which advocates on behalf of community gardens, and 596 Acres, which helps communities gain access to vacant public land, to get the city to dedicate even larger amounts of land for food production. Rather than simply focusing on preserving existing gardens, these organizations have urged community control of vacant land, land control as a political right, and the overall expansion of urban agriculture. In the case of allocating land for urban agriculture, practices and policy making have coevolved, with actions of gardeners, advocates, and entrepreneurs helping to shape policy and public plans, providing new opportunities for the creation of gardens and farms.

SUPPORTING DIVERSE URBAN-AGRICULTURE PRACTICES

Just as the actions of policy makers, urban gardeners and farmers, and urban-agriculture advocates have helped make more land available for food production, three different types of innovative agricultural practices have broadened the notion of producing food in New York City: (1) practices that span different spatial dimensions, including rooftop farms, building-integrated agriculture, and distributed plots of urban farmland; (2) innovations in permanency, or temporary and movable farms and gardens; and (3) an expanded notion of farming practice that includes addressing the racial, gender, ethnic, and class disparities that make urban food systems unjust.

Spatial innovations

New York City is considered a leader in rooftop agriculture, an example of an innovative practice of turning novel spaces into farms and greenhouses. Over the last few years, numerous rooftop-agriculture projects have been developed in New York City, where there is substantial potential for continued expansion because an estimated three thousand acres of flat rooftop space exist on buildings with the structural integrity to carry the weight of a rooftop farm (Ackerman 2011). The actors that have taken to these roofs are diverse and their projects offer multifunctional benefits. Rooftop spaces have been converted to therapeutic gardens in an assisted-living facility for formerly homeless adults; housing developers have built rooftop greenhouses into their affordable housing projects for job creation and health promotion; entrepreneurs have started for-profit rooftop farms; and grocers and restaurants are growing food on their roofs for their customers.

In the case of rooftop agriculture, entrepreneurship and innovative practices preceded policy making. Because of its popularity and potential as an economic development opportunity for the city, plans and policies followed the growth of this innovative practice of food production and now aim to support and expand it. For example, shortly after the creation of Brooklyn Grange rooftop farm, a nearly one-acre farm on a roof in Long Island City, Queens, and proposals for hydroponic rooftop greenhouses in Brooklyn, the City Council adopted Local Law 49 of 2011, which amended the building code by adding greenhouses to the list of rooftop structures (such as water tanks and ventilation equipment) that do not count toward building height limits (provided that the greenhouses occupy less than one-third of a roof's area). In addition, an amendment to New York City's zoning text was approved in 2012 that excluded rooftop greenhouses atop commercial buildings from the lot's floor area and height limits, and public funds have subsidized rooftop greenhouses and farms. These measures taken by New York City have been justified in planning documents like FoodWorks as a means to put unused space to productive use; to create jobs and economic value; to provide roof insulation, thus reducing energy consumption; and to capture and absorb rain, thus reducing pollution from stormwater overflow.

Innovations in temporality

Innovative urban-agriculture practices have also treated the temporality of farming differently. Rather than thinking of farm sites as permanent, these projects view agriculture as less place dependent, with production moving through the city to unoccupied locations as the use of space evolves. Moveable agriculture is not a new phenomenon, as abandoned lots have been turned into gardens during previous periods of economic disinvestment. However, their contemporary occurrence no longer reflects a makeshift or "guerrilla" option but is being considered a strategy of property owners and developers. Some developers have embraced temporary urban agriculture as a way to put their sites to productive use while scheduled development projects are on hold.

In New York City, Riverpark Farm is an example of interim urban agriculture (Cohen, Reynolds, and Sanghvi 2012). The farm grows vegetables in milk cartons filled with soil for the adjacent restaurant of celebrity chef Tom Colicchio, who helped fund the project. Although Riverpark neither required government approvals nor stimulated overt changes to public policy, the project has contributed to reenvisioning farms as temporary uses of properties by making the idea of urban agriculture possible in many more spaces in the city; demonstrating the benefit of temporary urban agriculture as a residential amenity; and showcasing and testing the design concept of growing food on a production scale using little more than soil-filled milk cartons. The low-technology farming method has allowed for the easy movement of the "farm" (which has happened several times since its inception, as different areas of the development site underwent construction), required horticultural innovations, and in turn spawned a new company (Rooftop Seeds) that supplies seeds adapted to challenging urban growing conditions. This New York example started at the entrepreneurial level, but in cities like San Francisco, officials created a formal policy allowing developers to avoid having to renew development approvals by allowing interim farms on their temporarily undeveloped sites.

Social justice practice innovations

Many urban-agriculture practitioners have broader goals and objectives that guide their work than merely growing their own food. Although important on its own, food production is seen as a means to address other social, environmental, economic, and health problems facing communities, such as access to healthy and affordable food; intergenerational relations; the creation of safe spaces; empowerment and mobilization; education and training; job growth and local economic development; biodiversity and habitat improvement; stormwater management; and soil improvement (Cohen, Reynolds, and Sanghvi 2012). As such, a single urban-agriculture practice can contribute to multiple benefits, involve a number of individuals and institutions, and serve as a mechanism for social and political change through empowerment.

In New York, some practitioners explicitly use their urban-agriculture projects as a way to address structural oppression and the racial, gender, and class disparities that result in disinvestment, unemployment, poverty, poor health care, and insufficient access to healthy food (Reynolds and Cohen 2016). Activist urban farmers and gardeners use the spaces and activities related to food production to educate, empower, organize, and engage in political mobilization. For example, Farm School NYC offers educational programs focused on advancing social justice and building self-reliant communities; the farm Granja los Colibries in Brooklyn provides a space for recent immigrants to participate in and lead activities that resist cultural assimilation while teaching about food and health disparities; and La Finca del Sur, a farm in the South Bronx created for and operated by "women of color and their allies," has built its work around women's empowerment through the farm and garden site. These cases and other social justice–focused farms and gardens aim to influence public policy on urban agriculture as well as related policy processes, such as participatory budgeting, land-use planning, and the more equitable distribution of resources among urban-agriculture groups (Reynolds and Cohen 2016).

These examples underline how changing the meanings, competencies, and material dimensions of urban agriculture alters the mix of farming and gardening practices, which in turn influences plans and policies. Rooftop farms, temporary farms, and gardens focused on social justice change our understanding of what a farm is and can accomplish, while new infrastructure, technologies, skills, and knowledge supported by policies make these innovations possible. Rather than taking a narrow view of what urban agriculture is (e.g., community gardens with individual plots or urban farms cultivated by a nonprofit or for-profit business), a more expansive understanding of urban-agriculture practice illustrates the multiple and different advantages that diverse practices can bring to neighborhoods. These innovative practices, in turn, stimulate policies and programs that support and normalize these practices in a coevolutionary process.

INNOVATING GOVERNMENT PRACTICES

Governments (and other stakeholders) contribute to urban agriculture by providing material and financial resources, yet as cities have faced significant economic repercussions from the recession (including declining local tax bases, rising municipal costs, and reduced federal aid), some cities have been reluctant to commit to such investments. Nevertheless, in New York and elsewhere, innovative practices in government have coevolved with public policies to connect different agencies to support the expansion of urban agriculture, in an effort to achieve the multidimensional benefits that urban farms and gardens provide while addressing multiple public policy issues simultaneously. In New York City, government agencies have changed practices to incorporate farms and gardens into their operations and development projects in support of their missions. Two examples include integrating agriculture into new residential buildings and incorporating urban agriculture into the city's green infrastructure program.

BUILDING-INTEGRATED AGRICULTURE

Urban agriculture can be integrated into the built landscape at multiple scales, and some cities have strategically created incentives for developers to provide growing spaces in their projects. In New York City, two public agencies—the Department of Housing Preservation and Development (HPD), which develops affordable housing, and the New York City Housing Authority (NYCHA), which runs the city's public housing projects and is also developing new affordable housing—have encouraged developers to design new housing with community gardens, production greenhouses, and even a rooftop apple orchard. They have done this through various practice innovations: by writing requests for proposals to encourage innovative designs to support the health of building residents; by funding the added costs of greenhouses and other agriculture-related building infrastructure; and by supporting nonprofit organizations to provide programming so that the agricultural spaces are used productively.

An example in the case of HPD is the request the agency issued for proposals for a 202-apartment affordable housing project in the South Bronx that required respondents to consider incorporating access to nutritious food, physical fitness, and places for social gathering in their proposals. The winning project was a design that included a small apple orchard, rooftop community gardens, and a community kitchen to teach food preparation skills (City of New York 2006). The resulting building, called Via Verde, was made possible by changes in practice within the agency to create an incentive for this type of design, which was in turn encouraged by policy in the form of healthy building design guidelines developed by the city's Department of

Design and Construction, as well as the practices of urban farmers and gardeners who have raised the visibility and popularity of urban agriculture in this low-income community. Via Verde and the efforts by HPD have subsequently resulted in additional building-integrated urban-agriculture projects in the South Bronx and beyond.

The NYCHA has also changed its practices to promote building-integrated urban agriculture. In one exemplary project, it sold a parcel of land on the grounds of a Bronx public housing project to a developer for the construction of Arbor House, a 124-unit affordable housing development. The developer secured funds from city officials, including the Bronx Borough president, to incorporate a hydroponic rooftop greenhouse to grow produce on a commercial basis for the surrounding low-income community (US HUD 2011). Forty percent of the produce grown is to be made available to local residents, schools, hospitals, and markets. As a result of its experience in this project, the NYCHA collaborated with the same developer to build a 364-unit mixed public and affordable housing project in the low-income Ocean Hill–Brownsville neighborhood of Brooklyn, which will include a supermarket and a rooftop greenhouse. In other instances, the NYCHA has collaborated with nonprofit organizations and other city agencies in creating farms that produce fresh vegetables and provide education, job training, and a space and activities that allow public-housing residents to socialize.

The practices of nonprofit organizations, developers, and city officials helped create these new urban-agriculture projects, which demonstrated to agencies like HPD and NYCHA that it is structurally and financially feasible, beneficial, and popular to design urban agriculture into public projects. The shift in practices has contributed to broader food policies, such as the city's anti-obesity plan, which encourages cooperation among multiple agencies to address diet-related health problems through urban agriculture and healthy building design.

URBAN AGRICULTURE AS STORMWATER MANAGEMENT PRACTICE

In another coevolution of practice and policy, New York City's Department of Environmental Protection (DEP) is funding the creation of some urban farms and gardens through a program to address the water pollution caused by combined sewer overflow (the discharge of untreated stormwater and sewage when it rains). Instead of investing solely in a traditional control method focused on conventional "gray" infrastructure (e.g., expanding water pollution treatment facilities and increasing the diameter of sewage pipes), the DEP opted for lower-tech landscape design interventions ("green" infrastructure) that increase the permeability of the cityscape through parks, landscaped median strips on roadways, and permeable pavement. Urban farms and gardens are a form of green infrastructure that have been funded through a Green Infrastructure Grant Program in which the DEP funds private property owners and organizations to build landscapes designed to capture and retain a minimum of one inch of stormwater from the impervious tributary area (Cohen and Wijsman 2014).

Urban-agriculture practitioners identified the grant program as an opportunity to convince the agency to consider farms and gardens a green infrastructure and have applied for funds. Since 2011, New York City has provided over $1.3 million in funding through this program to four urban-agriculture projects (with two more under consideration), including a one-acre commercial rooftop farm, Brooklyn Grange. However, although the DEP views urban agriculture or edible landscaping as a positive feature of a project proposal because of the co-benefits of food production, the agency's focus is on the ability of a project to reduce stormwater, not the production of fresh vegetables or any other benefits of urban agriculture. This puts the onus on the city's urban-agriculture community to propose new farming projects for funding under this program and to evolve its practices (e.g., entering into long-term leases with property owners, demonstrating organizational capacity, agreeing to track stormwater absorption over the life of the project) to qualify for these funds (Cohen and Wijsman 2014).

This example illustrates that innovations in practice, such as the adoption by the DEP of new approaches to stormwater mitigation or new techniques by farmers such as growing food atop asphalt roofs, contribute to changes in agency policies and programs, and that these policies, often embodied in a strategy like the department's green infrastructure plan, help support and replicate these innovative practices. Plans themselves do not lead to changed practices, nor do innovative practices emerge in a vacuum without policies and a plan to provide funding and other forms of technical and material support. Policies help change practices but depend on the existence of innovative practices to gain the political traction needed for adoption and implementation. Practitioners and policy makers "enroll" new practitioners, like engineers at the DEP or housing finance specialists at HPD, in support of urban-agriculture practices, broadening the scope of policies supporting urban agriculture and thus solidifying support for the practice.

DISCUSSION

Changes in urban-agriculture practices, planning, and policy making occur simultaneously and in reaction and response to each other. For example, the efforts of farmers and gardeners to take over vacant spaces to grow food has helped build a case for incorporating food production into the urban environment, while new local laws have created a process for doing so, thus legitimizing alternative, bottom-up methods of securing control of city-owned property. Novel farming and gardening practices such as rooftop farms, temporary farms, and farms and gardens dedicated to social transformation have shaped plans and policies by expanding the scope of possibilities and meanings of urban agriculture beyond mere food production. The same coevolutionary dynamic occurs within government, as agency officials engage in new practices that expand the boundaries of their work. These agency practices, such as housing agencies integrating urban agriculture into new residential projects, break down administrative silos within city government, broadening the range of actors in government who view urban-agriculture planning as within their purview, and enabling the practices of varied agencies to be incorporated into plans and policies.

The challenge for creating a resilient and equitable food system is imagining and supporting desired versions of food production practices. Since this requires influencing the practices in which people engage, specifically how they grow (and procure, process, and discard) food, planners and policy makers must understand the competencies, materials, and meanings that influence food production practices. The interplay of practice and policy, and understanding how policy reconfigures the material elements, meanings, and knowledge that engage people in practices, can help cities influence the technologies, routines, and forms of everyday practices that embed urban food production in city life.

The focus on practice has important implications for planners and policy makers involved in supporting and expanding urban agriculture, and for food systems planning generally. As the New York City examples show, the efforts needed to advance urban agriculture are as much about identifying, guiding, and supporting new practices, including deviant practices that are considered beneficial, through the creation of policies and programs, as they are about producing plans with a priori goals, objectives, and strategies. If the aim is to understand and evolve the range of urban-agriculture practices that shape our cities, planning researchers (both academics and practitioner-researchers) must focus on two aspects of practice. First, planners and policy makers must trace the emergence, disappearance, or transformation of relevant practices, and the cultural, material, and technical dimensions of practices that lead them to be replicated and adopted, to fail, or to change. This means examining as many of the forces that influence practices as possible to understand whether and how to support, replicate, expand, change, or stop them. Second, they must measure the impacts (on the environment, social equity, and urban economy) of practices, including the anticipated impacts if practices were to become established more broadly, or if they were to cease, recognizing that it is difficult to quantify the causal relationships between a practice that changes a complex system and specific outcomes.

DISCUSSION QUESTIONS

1. Who is it that makes (or made, in the case of New York City) urban agriculture a norm in the city—"guerrilla gardeners," or government programs like GreenThumb?
2. Was it the popularity of the community garden or the garden's implementation within city government that was the greater cause of Mayor Giuliani's defeat in attempting to develop the land?
3. Examine the role of New York City government in urban agriculture. What parts of its involvement help urban agriculture? Are there any parts or actions that are not helpful? In the future, how can government better facilitate urban agriculture?

4. What advantages exist in encouraging interim urban agriculture as described in the "Innovations in temporality" section? What limitations or issues exist with this concept?
5. New York saw great success in promoting urban agriculture through the development of new housing. Is this feasible and effective for every community? Explain your viewpoint.
6. Imagine you are a planner urging another city to promote urban agriculture in the ways that New York City does. What policies in particular should other cities pursue and why? If there are policies that cities should not pursue, also elaborate on those.
7. Cohen and Wijsman point to social justice and equity as a popular instigator of urban-agriculture projects. What instigates policy makers to pursue urban agriculture?

CHAPTER 14

Urban Agriculture and Health

What Is Known, What Is Possible?

BENJAMIN W. CHRISINGER AND SHEILA GOLDEN

INTRODUCTION

Urban agriculture has been celebrated for its ability to increase access to and promote consumption of healthy, fresh foods, increase nutritional knowledge, provide restorative spaces, and build social capital in communities (Hodgson, Campbell, and Bailkey 2011). In the last five years, many metropolitan areas across the United States have revised zoning and land-use policies to accommodate urban agriculture (Goldstein et al. 2011; Hodgson 2012; Hendrickson and Porth 2012). Many of these efforts are inspired by a growing body of literature and research that describes beneficial health impacts of urban agriculture. To further develop our understanding of the connections between urban agriculture and health and inform future efforts to improve health with urban agriculture, a focused and practice-oriented research agenda is critical. New research can build compelling datasets and offer useful insights that draw support from public health policy makers and funders.

Researching the health impacts of urban agriculture is a difficult task, particularly considering the financial and time limitations of many urban-agriculture programs. Nonetheless, spending time and resources on evaluation and research can lend greater weight to the larger urban-agriculture movement and may also help increase the competitiveness of grant applications for individual projects. Recognizing this, the chapter is divided into two parts: first, a summary of health benefits linked to urban agriculture found through a comprehensive literature review, and second, ways to think about measuring possible health impacts going forward.

We begin by reviewing research that has uncovered a variety of health measures and outcomes associated with urban agriculture, including increased fruit and vegetable consumption as well as improvements to diet, physical activity, psychosocial state, body mass index (BMI), and social cohesion. To broaden the scope of this review, we also present studies documenting connections between health and food markets, such as farmers markets and community-supported agriculture (CSA) programs.

The chapter concludes by suggesting tools and frameworks that practitioners can use to tailor evaluations, design health interventions, or conduct general research on the health benefits of urban agriculture. We suggest two different approaches to this research: the first asks how urban agriculture provides access and opportunities that improve health outcomes, and the other considers how urban agriculture might influence healthful attitudes and abilities. For each approach, we offer tools and examples of effective models. We focus broadly on key theories and concepts about the connections between urban agriculture and health, in the hope that readers will apply these ideas in the more detailed case studies and examples that appear in other chapters.

STATE OF THE EVIDENCE: HEALTH EFFECTS OF URBAN AGRICULTURE

Are urban-agriculture participants or beneficiaries healthier?

Research documenting the health effects of urban agriculture generally falls into one of four categories: dietary habits, physical activity, physiological outcomes, and psychosocial outcomes.

Generally speaking, we can measure health outcomes in one of two ways: direct measures and indirect measures. Direct measures are most often characteristics that are observed or measured in person, such as BMI, heart rate, or cholesterol levels. Some of these measures are easy to collect, like height and weight (for BMI), while others can be very intrusive and require the collection of medical samples (like a blood draw), dramatically raising the cost and complexity of a study. Because certain health outcomes are difficult to measure directly, researchers frequently choose indirect, observational, or self-reported measures as alternatives.

Indirect measures might include dietary recalls, measures of fruit and vegetable consumption, or questionnaires to assess average levels of physical activity. These types of measures can produce equally high-quality and convincing data, as long as researchers are aware of how these data are context-specific representations of health (i.e., *self-reported* data might depend on how comfortable an individual is with sharing or how well they remember that information; *directly measured* weight is as accurate as the scale used).

Several resources and tools currently provide more detail and information on researched health impacts of urban agriculture. Practitioners who are looking for existing research might find these tools useful.

1. The Community Food Security Coalition's North American Initiative on Urban Agriculture published a summary of research on health benefits of urban agriculture (Bellows, Brown, and Smit 2005) followed by a literature review of research suggesting nutrition implications of urban farmers markets and community gardens (McCormack et al. 2010).
2. The University of California's Division of Agriculture and Natural Resources published an updated literature review on social, health, and economic impacts of urban agriculture (Golden 2013c). This literature review is accompanied by an annotated bibliography of all the cited literature (Golden 2013a), as well as an "At a Glance" spreadsheet that attaches articles and reports to their specific impacts (Golden 2013b).
3. The Five Borough Farm Project in New York City published an annotated bibliography that is organized by metrics (Sanghvi 2012) and lists research that links positive benefits to urban agriculture. The project's website includes graphics that outline metrics frameworks

and tools to design research and evaluation of urban-agriculture benefits (www.fiveboroughfarm.org).

DIETARY HABITS

The literature describing connections between urban agriculture and dietary practices is sparse, and most studies are single cases rather than systematic assessments (McCormack et al. 2010). The research described here offers a brief picture of what has been documented.

Increased consumption of fruits and vegetables

Perhaps the most common of the dietary evaluations are measures of fruit and vegetable consumption. In five studies that used surveys, individuals who participated or had family members who participated in community gardens reported more servings of fruits and vegetables than control groups who did not garden (Litt et al. 2011; Wakefield et al. 2007; Alaimo et al. 2008; Lackey 1998; Twiss et al. 2003; Blair, Giesecke, and Sherman 1991). Two reviews of garden-based youth nutrition programs operating in urban areas also found promising reports of increased fruit and vegetable consumption among participants (Twiss et al. 2003; Robinson-O'Brien, Story, and Heim 2009).

Urban-agriculture marketing opportunities such as farmers markets and CSAs are also associated with more healthful food consumption. Neighborhoods with farmers markets had higher fruit and vegetable consumption rates among people of color (Park et al. 2011). One study found that WIC participants who received coupons and shopped at farmers markets consumed more vegetables than those who shopped at grocery stores or did not receive coupons (Herman et al. 2008). Studies on CSA member consumption found that people belonging to CSAs used most of their issued produce (Landis et al. 2010) and were likely to consume greater amounts and more varieties of fruits and vegetables (Kerton and Sinclair 2009; Landis et al. 2010; Sharp, Imerman, and Peters 2002).

Less consumption of unhealthy foods

Beyond the promotion of fruits and vegetables, some have documented decreases in urban-agriculture participants' consumption of unhealthy foods.

One Philadelphia-based survey found that urban gardeners reported consuming less dairy and sweet foods and beverages than nongardening peers (Blair, Giesecke, and Sherman 1991). Also, youth involved in community garden programs discussed eating less unhealthy food, such as candy, as a result of their participation in a community garden program in Flint, Michigan (Ober Allen et al. 2008). Thus, there is some indication that participation in urban agriculture can promote improved diet beyond fruits and vegetables alone.

Improved perceptions of fruits and vegetables

Recognizing that having fruits and vegetables available does not necessarily translate to their consumption, some studies have attempted to gauge the intermediate effect of urban agriculture on attitudes toward eating fruits and vegetables. For example, youth-based studies have documented how garden-based nutrition programs can improve a child's perceptions of fruits and vegetables, possibly increasing the likelihood of consumption (Gatto et al. 2012; Robinson-O'Brien, Story, and Heim 2009; Ober Allen et al. 2008).

PHYSICAL ACTIVITY

Regular physical activity is well understood to have positive health effects, including reduced risk of many serious chronic diseases like diabetes, cancer, and obesity (Warburton, Nicol, and Bredin 2006). Some medical studies have specifically referenced gardening as a potentially useful form of physical activity, akin to brisk walking or bicycling (Magnus, Matroos, and Strackee 1979; Wannamethee and Shaper 2001). In several case studies, participants stated that a major motivation or outcome of participating in urban agriculture was increased physical activity (Armstrong 2000a; Twiss et al. 2003; Wakefield et al. 2007; Saldivar-Tanaka and Krasny 2004).

PHYSIOLOGICAL OUTCOMES

Urban agriculture—specifically urban gardening—has been the chosen method of intervention in several health promotion efforts. These projects deliberately measured specific health indicators, depending on the aims of the intervention;

for instance, one study documented the positive effects of gardening in the active management of diabetes (Weltin and Lavin 2012). Another study found positive physical effects of gardening among an elderly population, using a validated clinical instrument (Short Form 36 Health Survey) to consider a range of outcomes (Park, Shoemaker, and Haub 2009). While both of these projects were tailored efforts to improve health among specific populations, the documented health effects may resonate beyond these communities.

Healthy BMI has also been suggested as a possible benefit of urban-agriculture participation, based on a study of community gardeners in Salt Lake City, Utah, which found that gardeners had significantly lower BMI than their nongardening neighbors (Zick et al. 2013). Given that weight status is determined by many different factors, researchers in this study recruited three separate control groups to account for other potentially influential variables, including genetics, nutritional access, and neighborhood environment.

PSYCHOSOCIAL OUTCOMES

Measures of mental state, interpersonal relationships, and social cohesion contribute to our understanding of psychosocial well-being. Much of this knowledge comes from qualitative case studies, surveys, ethnographies (long-term observational research), and in-depth interviews to gather data. Psychosocial health measures are intrusive and difficult to quantify by direct measures (e.g., brain activity); thus many studies employ indirect, self-reported estimates of stress, happiness, or friendship. Studies that employ these self-reported metrics must take care to control for subjectivity; do all study participants agree on definitions and degrees of stress, happiness, or friendship (Kahneman and Krueger 2006)? Despite these logistical hurdles, several studies observed positive social outcomes for individuals and communities that participated in various forms of urban agriculture, including increased social capital, benefits for seniors, and improved well-being through the creation of safe and green spaces.

Increased social capital

Some studies found that urban agriculture—particularly community gardening—can promote civic engagement and social connection between individuals.

Several urban-agriculture projects provided immigrants with an opportunity to network with other immigrants and created shared opportunities with nonimmigrant residents by growing, trading, and often selling their produce (Krasny and Doyle 2002; Beckie and Bogdan 2010). Community gardens were cited as important spaces for gathering and socializing (Patel 1991; Saldivar-Tanaka and Krasny 2004; Teig et al. 2009; Wakefield et al. 2007). Many articles analyzed how these interactions involved decision-making and planning processes that required consensus, making community gardens important places for fostering democratic values and citizen engagement (Glover, Shinew, and Parry 2005; Mendes et al. 2008; Patel 1991; Teig et al. 2009; Travaline and Hunold 2010). For urban farms and businesses, researchers found that participants cited self-determination, self-reliance, and activism as major impacts (Bradley and Galt 2013; Colasanti, Litjens, and Hamm 2010; McClintock 2014; White 2010). Many project participants discussed improved self-esteem and pride in their work (Feenstra, McGrew, and Campbell 1999; Bradley and Galt 2013). In both community gardens and urban farms, the advocacy and coalition building needed to overcome structural barriers of zoning, land-use conflicts, and resource shortages created "networked movements" (Mendes et al. 2008) and a new generation of activists and engaged citizens (Levkoe 2006; Sumner, Mair, and Nelson 2010; White 2010).

Farmers markets were also discussed as places for gathering and fostering community. However, a number of articles discussed barriers, such as lack of affordability and culturally appropriate food and space, that excluded low-income and minority residents (Fisher 1999; Suarez-Balcazar 2006).

Cross-generational integration and senior well-being

Urban agriculture is also a way to promote senior well-being and cross-generation sharing between youth and seniors. Since the majority of community gardeners are seniors (Armstrong 2000b; Patel 1991; Schukoske 2000; Teig et al. 2009), these gardens are an ideal venue for seniors to pass on knowledge and work with youth. In one study, seniors claimed that garden spaces sometimes helped them transition from home ownership to senior homes and higher-density living (Armstrong 2000b). Other studies suggest positive mental and social benefits of gardening activities for seniors (Austin, Johnston, and Morgan 2006; Park, Shoemaker, and Haub 2009).

Creating safe places and reducing blight

Community gardens and urban farms create safe spaces to recreate and improve the physical space of the neighborhood. Participants said that gardens and farms beautified their neighborhoods and employed and benefited residents, which in turn created more local pride and attachment to the space (Bradley and Galt 2013; Ober Allen et al. 2008; Alaimo et al. 2008). Participants expressed that the presence of the farms and gardens helped decrease vandalism and criminal activity and increased safety (Bradley and Galt 2013; Ober Allen et al. 2008; Teig et al. 2009). Community gardens, in particular, were cited as a place where people built trust and rapport (Teig et al. 2009; Kingsley, Townsend, and Henderson-Wilson 2009), cultivating general well-being. Also, studies have documented the prosocial effects of greening vacant lots (Branas et al. 2011; Garvin, Cannuscio, and Branas 2013); assuming that urban-agriculture operations effectively emulate these greening procedures, they may also yield similar psychosocial benefits.

DETRIMENTAL HEALTH EFFECTS?

It is worth noting that researchers have also examined urban agriculture as a potential health risk, largely related to the use of contaminated sites, water sources, or animal wastes (Flynn 1999). While possible detrimental health effects of urban agriculture are not extensively explored in American cities, the legacy of early planners and public health practitioners continues to influence urban-agriculture policies. The efforts of these reformers largely pushed agriculture, especially practices involving animals, outside city boundaries and cited public health as a primary concern (Brinkley and Vitiello 2013). As planners seek to amend or update zoning policies to expand urban agriculture, they should keep an eye to the historical policy precedents in their own cities.

CONTEXT AND COMPLEXITY

Even if we can document direct health benefits from urban-agriculture activities, we should be careful not to overgeneralize the results. Given a set of particular pathways and health outcomes, what other contextual variables could also matter? For instance, if a study measures gardeners who are high-income retirees, it is unlikely for them to cite a lack of time or money as reasons for not consuming fruits or vegetables. This is not to say that low-income workers are unable to participate in urban agriculture; rather, they face different constraints that must be accounted for in statistical analyses to prevent biased results. In this example of trying to estimate the health effects of urban gardens, income and employment status are likely to be influential determinants of health, perhaps eclipsing possible effects of gardening.

A useful method for considering this complexity is to map the causal pathways, even speculatively, between urban-agriculture features and health outcomes. Causal pathways (fig. 17) connect social, genetic, and behavioral determinants of health to physical outcomes. The following section will outline several possible causal pathways between urban agriculture and human health that planners, policy makers, and researchers could measure in their own communities.

CAUSAL PATHWAYS BETWEEN URBAN AGRICULTURE AND HEALTH

How might urban agriculture cause participants or beneficiaries to become healthier?

Although more studies and evaluations are documenting the health benefits of urban agriculture, there are still relatively few that employ direct measures of health effects. The small number of completed clinical studies illustrates the cost and complexity of the research design required to make causal claims. When program designers and planners consider the *possible* health effects of urban agriculture, they should try to form reasonable expectations and targeted research designs.

To turn again to the example of community gardeners with lower BMI, we can form reasonable expectations by better understanding exactly *how* the practice of gardening promotes lower weight status. Body weight is understood as a product of diet and physical activity; thus, the primary unanswered question is whether gardening more strongly affects diet, physical activity, or both in some reciprocal fashion. Secondary questions would center on exactly

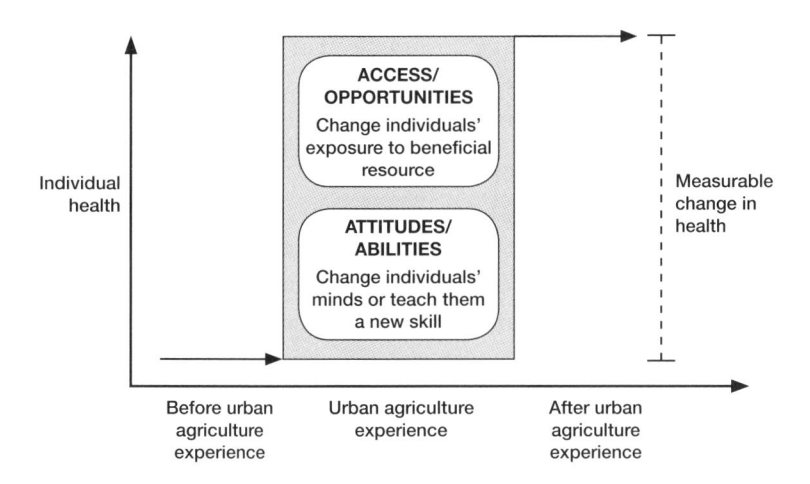

Figure 17. Causal pathways of health.

how gardening supports a better diet or levels of physical activity, how this varies by type of gardening and goals for the gardener, and thus, whether or not this is likely true for all gardeners.

Causal pathways of health

Two groups of causal pathways characterize the exact ways health could be affected by urban agriculture (see fig. 17 for a diagram of these concepts). The first group, *access and opportunities*, might improve health because they afford individuals additional exposure to positive environments or resources (or, alternatively, reduced exposure to negative ones). The second group, *attitudes and abilities*, could improve health by changing an individual's mindset through education or skill development. The distinctions between groups are important to consider in efforts to design, implement, and evaluate projects to improve health. The Five Borough Farm Project, a project of the Design Trust for Open Space in New York City, provides a useful metrics framework and data collection tools that reflect many of the pathways (www.fiveboroughfarm.org). Health Impact Assessment (HIA) is another potentially useful process that offers a systematic method for considering health effects (www.healthimpactproject.org). These resources can be useful in helping practitioners track and evaluate urban agriculture's possible health impacts. We also provide other examples of methods and freely available resources for measuring these different types of causal pathways.

Access and opportunities

Causal pathways in this group help answer the question, How does urban agriculture make healthy lifestyles *more accessible* to individuals or *create opportunities* for them to execute healthy behaviors? Generally speaking, these pathways are easier to measure and might involve tools like geographic information systems, open-source mapping tools, or price and availability surveys. Place-based interventions to improve physical access may also be more straightforward interventions for planners and policy makers, especially compared to the more complicated task of changing human behavior. Interventions that rely on these types of mechanisms are considered to be place-based strategies; in essence, they are efforts to "level the playing field." Let us consider several ways urban agriculture could improve access and opportunities for health.

Lower physical and economic barriers to healthy options   Many point to the high cost of fruits and vegetables compared to cheaper, highly processed foods as a contributor to diet-related health disparities. Similarly, the previous decade of food desert research has also cited low physical accessibility to healthy options as problematic for community health. Urban agriculture may be able to lower the direct costs for these items and bring them closer to disadvantaged communities, many of which cannot economically support large food retailers. Studies could map the availability of healthy food retailers using freely available resources like Community Commons (www.communitycommons.org), overlaying other important neighborhood characteristics, like household income or family size (see also a more nuanced definition of access in chapter 12 of this volume). The local cost of healthy food could also be measured with straightforward in-store audits (for examples, see the National Cancer Institute's online measures database, http://appliedresearch.cancer.gov/mfe/instruments).

Exposure to green space   Spaces where urban agriculture is practiced may also afford community members opportunities to interact with nature. This exposure could occur in discrete events, such as an afternoon of gardening, or more indirectly over time, such as an urban farm operating across the street. These differences in exposure determine the exact kinds of access and opportunities that urban agriculture may provide. Researchers can either consider how much exposure individuals have from home with an area-based measure (e.g., square feet of green space on a given block), or engage with community members to more carefully consider exposure (e.g., in-person interviews to determine frequent walking routes). A variety of quantitative and qualitative measures were used by Branas et al. (2011) and Garvin, Cannuscio, and Branas (2013) to consider the effects of a Philadelphia vacant lot greening program, and these studies might provide urban-agriculture researchers with useful methods or ideas.

Space for physical activity and social interaction   A lack of safe, quality environments for physical recreation and social interaction has been cited as a possible contributor to low levels of recreation and connectivity among disadvantaged neighborhoods. As discussed in the literature, urban agriculture can provide clean and safe spaces where participants can engage in various activities, including exercise, collaboration, and interaction. In order to understand the unique recreational preferences or unique circumstances in each community, it is compelling to have community members identify these spaces and characterize them. Participatory action research and community reporting are great ways to collect relevant and more-accurate data. Open-source mapping programs such as Google Maps can be used as an accessible and easy database for residents to enter, share, and collaboratively interpret data.

Attitudes and abilities

Causal pathways in this group help answer the question, How does urban agriculture *change individuals' minds* about healthy lifestyles or *make them more able* to execute healthy behaviors? Measuring these pathways may be more complicated, especially given the need for pre- and post-exposure data collection to measure true change in attitudes or abilities. Nonetheless, surveys or observational studies can provide a good start, even if individuals are simply self-reporting changes postexposure (see the measures registry of the National Collaborative on Childhood Obesity Research for a variety of examples, www.tools.nccor.org/measures). Interventions using these types of mechanisms are typically considered to be people based and often focus on capacity building and lowering mental barriers to changing behavior. Urban agriculture could influence individuals' attitudes and abilities in a number of different ways.

Increased willingness to try healthy options Both children and adults who participate in urban-agriculture activities may be planting, harvesting, or selling previously unfamiliar fruits and vegetables. Participants may be more willing to try new foods they had a hand in producing, or curious neighbors could explore options they would not otherwise consider in a conventional retail setting. Measurements of this pathway could document how an individual's attitudes toward fruits and vegetables changed with exposure to urban agriculture. Farm to School programs often conduct evaluations that include tastings where students are given samples of fruits and vegetables (often grown in their school garden programs) and surveyed. Some of these evaluation strategies could be easily adapted for urban-agriculture programs. The National Farm to School Network has a great tool kit with resources for evaluating attitudes toward healthy options that includes survey prototypes and evaluation designs (Joshi and Azuma 2012).

Increased knowledge of healthy options By exposing participants and beneficiaries to different varieties of fruits and vegetables, urban agriculture can help build nutritional knowledge. For instance, urban-agriculture activities could provide tangible examples of recommended healthy foods, servings, or dishes. Individuals may learn how different types of foods can be used to manage or improve existing health conditions, like obesity or diabetes. Programs can measure this by using surveys and pre- and post-interviews. The Farm to School evaluation tool kit mentioned earlier provides survey samples that can direct evaluation design (Joshi and Azuma 2012).

Increased personal contact and social capital Through participation in urban-agriculture activities, individuals may develop marketable skills, self-motivation, and civic pride (see chapters 7 and 8 for examples from community case studies). It is also possible that the networks and associations formed through urban-agriculture projects could increase social connectivity and collective efficacy. By improving contact between neighborhood residents, urban agriculture could decrease any existing negative effects of isolation or alienation. Self-reported indicators (e.g., "Do you feel connected to your neighborhood?") and other indirect measures (e.g., participation in local elections, neighborhood associations, etc.) could help describe changes in this mechanism. As discussed before, using community-generated data and participatory action research strategies would create a compelling dataset.

Both pathways matter

In practice, most health interventions employ both types of causal pathways. For instance, numerous garden-based efforts will attempt to increase access and opportunities to consume fruits and vegetables while also actively promoting these new foods through educational programs like cooking classes (see also the numerous examples in chapter 16 of this volume). These hybrid interventions could quite possibly employ pathways not considered in this chapter, adapting to meet the needs of a specific community or population. As program designers build evidence for their interventions, it is worth considering what types of pathways are at play, since they may require different types of tools to be appropriately measured.

CONCLUSION

While the breadth and depth of existing literature documenting the health effects of urban agriculture leave room for growth, the renewed interest by planning researchers and practitioners in both urban agriculture and public health holds promise for further investigations. Similarly, calls for politicians and public health officials to develop evidence-based health policies increase the likelihood for new cases to be added to this body of research. With added context and evidence, advocates for urban agriculture will be better able to accurately estimate the health outcomes of their efforts.

Though we are likely to see more research connecting urban agriculture and health, complex social, political, economic, and temporal factors typically make it impractical to conduct studies capable of making truly causal claims. However, planning practitioners and researchers can still contribute to a rich local understanding of health impacts by measuring the causal pathways of health (access and opportunities, attitudes and abilities) as a meaningful practical alternative to clinical studies.

This chapter has attempted to briefly describe research that has found connections between forms of urban agriculture and human health, while also providing a framework for considering causal pathways that could be considered in future programs and evaluations. It should remind readers of real-world health efforts in their own communities, where it is likely that a number of pathways are at play, all possibly influencing individual and community health. Going forward, these tools can offer a basis for creating additional knowledge and reasonable expectations about urban agriculture and health.

DISCUSSION QUESTIONS

1. Identify different types of evidence that are used to connect urban agriculture and health. Would different types of evidence be more/less appropriate if you were trying to implement an urban-agriculture ordinance versus design a diabetes-prevention program?
2. To what degree are there measurable positive effects of urban agriculture on people who are exposed to, but not necessarily participants in, urban-agriculture activities?
3. What factors could improve access to and use of farmers markets among low-income or minority communities?
4. How could mapping technologies be used to engage urban-agriculture participants? What types of spatial data might be useful for exploring community context?
5. As food system planners and health officials try to expand future collaborations, what challenges and opportunities might they encounter?

CHAPTER 15

More Than the Sum of Their Parts

An Exploration of the Connective and Facilitative Functions of Food Policy Councils

LINDSEY DAY FARNSWORTH

INTRODUCTION

Since the first food policy council was established in Knoxville, Tennessee, in 1982, over two hundred state and local food policy councils have been initiated across the United States. Much of this growth is very recent; there was a nearly seven-fold increase in the number of food policy councils in the United States between 2005 and 2015 (Center for a Livable Future 2015). The proliferation of the food policy council suggests that this organizational model is perceived to fill a critical niche in community and regional food systems—but which?

Inherently complex and highly interconnected, the food system has proven to be a challenge to coordinate and administer. Food policy takes a variety of forms and pertains to all phases of the food system, from production to waste. As a result, numerous local, state, and federal departments—ranging from Health and Human Services, to Agriculture, to Economic Development and Planning, to Parks and Recreation—oversee narrow subsets of food-related programs and policies.

The community and regional food systems literature has documented a variety of problems resulting from this decentralization and has highlighted issues associated with conflicting normative objectives and overall lack of alignment across food policies (Barling, Lang, and Caraher 2002; Connelly, Markey, and Roseland 2011; Dahlberg et al. 2002; Koc et al. 2008; Muller et al. 2009; Pothukuchi and Kaufman 1999; Wekerle 2004). Food policy councils can assume a variety of forms and functions, but their central purpose is to "provide a forum for diverse stakeholders to come together and address common concerns about food policy, including topics such as food security, farm policy, food regulations, environmental impacts, health, and nutrition" (Broad Leib 2012, 1). As such, food policy councils have emerged as a popular organizational strategy for improving communication, regulatory alignment, and problem solving across different agencies and sectors that shape our food system.

Despite their growing popularity, significant questions remain as to whether food policy councils will be able realize their potential. Concerns regarding food policy councils' overall lack of staff capacity, funding, and political authority permeate the literature. For example, according to Harper et al. (2009), "The vast majority of food policy councils have either no staff at all or only one part-time staff person, relying instead on volunteers or on restricted amounts of staff time from city, county or state employees assigned to the council in addition to their usual government duties" (3). This has already contributed to the dissolution of several food policy councils. Further, food policy councils that are not seated in government have been criticized for competing with member organizations for grant funding and thus detracting from, rather than enhancing, collaboration. There is also meager documentation of the impact of food policy councils. Harper et al. (2009) note, "We were unable to quantitatively demonstrate the impact of Food Policy Councils on food access, food policy, public health, or economic development due to a lack of data or evaluation procedures within individual councils, despite numerous success stories" (7). Finally, some have suggested that food policy councils have inherently limited efficacy because they are fundamentally neoliberal entities that were never intended to transform the current food system (Alkon and Agyeman 2011; McClintock 2014). Nevertheless, food policy councils continue experimenting with a range of organizational strategies to maximize their impact and demonstrate their worth. In the absence of full-time staff and stable sources of funding, the expertise and political and social capital of their organizational members are arguably food policy councils' greatest assets. As such, it is critical that food policy councils understand their members' self-interest and motivations for participating. In fact, food policy councils may find it fruitful to incorporate such considerations into their structures and processes to ensure their own success.

This chapter highlights initial findings from exploratory research conducted in conjunction with the University of Wisconsin–Madison Community and Regional Food Systems Project to understand what motivates individuals to participate in food policy councils and how their goals are addressed (or not) through participation in food policy council activities. We illustrate the connective and facilitative contributions of food policy councils to the ongoing work of their members by drawing on key interviews from a quota sample of members from two food policy councils. We first outline motivations for participating in food policy councils and provide examples of successes, then highlight ongoing challenges faced by these councils, and conclude with a discussion of the implications of these findings for food policy council structure and process.

METHODS

This chapter draws on twelve interviews with members of two food policy councils conducted between October 2012 and March 2013. While council selection was based largely on a convenience sample, the councils represent distinct geographic regions, membership structures, and city sizes. Because of significant variation across local food policy councils, the jurisdictional and geographic scope of their work varies; the two councils highlighted in this study have an urban focus but engage with county-level partners and issues. One of the two food policy councils examined in this study has an open membership policy. The other is made up of appointed members but its working groups welcome participation by the public.

A quota sampling method was used to select key informants from each food policy council to ensure representation of multiple vantage points. To achieve this, food policy council coordinators were asked to identify council members representative of the following topic areas: food justice, emergency food, urban agriculture, farm to institution, food access, and policy. These categories were selected because they are more specific than categories such as food production, distribution, and consumption, allowing for a more meaningful comparison across cases while still representing common issues in food policy councils. It bears mention that all of the interviewees who participated in this study were employed by organizations working on agriculture, food, or nutrition issues and were able to attend food policy council meetings in a professional capacity.

Interviews were conducted by phone and transcribed. The specific councils and interviewees have been kept anonymous to allow for the inclusion and discussion of potentially controversial perspectives. Generalizations based on these exploratory data reflect observations that surfaced across interviews in both cities.

Although this chapter highlights quotes and examples from only two food policy councils, it is also informed by sensitizing information from interviews with members of other food policy councils, participant observation of multistakeholder food council or coalition meetings in six cities and five states, and my direct involvement in the Madison, Wisconsin, food policy council. In analyzing the transcripts on which this chapter is based, I first used a priori codes formulated following a literature review of food policy councils. Emergent codes were added to accommodate new concepts as they surfaced. Further research is needed to ascertain whether and to what extent the themes and lessons that emerged from this exploratory study are consistent with the experiences of other food policy councils.

MOTIVATIONS FOR PARTICIPATING IN FOOD POLICY COUNCILS

Regardless of whether interviewees were formally appointed to a council or simply chose to partake in one, their motivations for participating in food policy councils were typically one or a combination of the following objectives: to increase coordination with other organizations active in the metro food system, to advance specific goals or projects, and to network and acquire information about existing activities.

Increasing coordination across urban food initiatives

Desire for increased coordination across urban food initiatives may be in part a response to the growing visibility and prevalence of local food-related activities. Consider the increases in food system activity associated solely with growth in consumer demand for local food and increased spending to fight diet-related diseases. The number of farmers markets in the United States has more than tripled over the past fifteen years and now exceeds eight thousand (USDA, n.d.). According to the USDA, local food sales were estimated to have climbed from $4.8 billion in 2008 to $7 billion in 2011 (Packaged Facts 2007).

At the same time, philanthropic and federal funding to fight diet-related diseases has increased dramatically. In 2007, the Robert Wood Johnson Foundation pledged $500 million to reverse the childhood obesity epidemic by promoting physical activity and increasing the accessibility of healthy food in school and retail environments (Strom 2007). In addition, through the American Recovery and Reinvestment Act of 2009, the Centers for Disease Control and Prevention released $373 million in grant funds for local "communities to adopt and implement evidence-based policies to improve nutrition, increase physical activity, decrease overweight and obesity and tobacco use" (Robert Wood Johnson Foundation, n.d.). Given the recent rise in spending on food and nutrition activities at the local level, it is not surprising that even those with a history of involvement in food issues cannot keep up with all the activity.

In the following quote, the director of a farm-to-table organization speaks to how community food systems can be rife with innovation and activity but lack the coordination and cross-pollination across activities that are necessary to shift from piecemeal progress to systemic change:

> I think we needed a food council. There was a screaming need for coordination of efforts. [This town] is kind of like the Silicon Valley of urban agriculture right now. You know how Apple started in a garage? Tomorrow's Growing Power[1] is probably sitting in [a coffee shop] right now pecking away at the keyboard. . . . So lots of folks are trying different things out and when you have a lot of independent entities all working on related things, it's important to know what other people are up to.

In the words of another council member, food policy councils are a forum for people already working on local food issues to "become aware of how the different components [of the food system] play together . . . and to see how we might be more effective in our food system work by sharing information and ideas and looking for opportunities for collaboration."

Advancing specific goals or projects

Another common motivation for participating in food policy councils was to advance specific goals or projects, typically related to council members' professional responsibilities. For example, one council member worked at a statewide food buying club that was initiating a mobile market in a central city. She became involved in the food policy council in the early 2000s out of general interest in improving communication and coordination, but she was especially motivated by a desire to meet new partners who could help identify specific sites in which to locate the urban mobile market.

Another council member, with a background in community economic development, learned about his local food policy council after contacting the Office of the Mayor to explore ways the city could be more supportive of food carts and street vending as a means of fostering small business development. In this instance, the food policy council provided an outlet for him to advance his work on street food in a context in which he could work with city staffers and other partners. At the same time, it enabled the Office of the Mayor to respond to his inquiry by redirecting him to a forum that was better equipped to support and develop his interest. While this individual initially became involved in the food policy council through its working group on street food, he has since become an appointed council member.

Networking and acquiring information about existing activities

Finally, and not surprisingly, early-career professionals and individuals who are newer to urban food issues often sought an information clearinghouse on local food system activities. They were more interested in informal networking opportunities, while seasoned professionals and activists tended to come to food policy councils with more focused objectives. The discrepancies between the needs and objectives of newer and more experienced activists and professionals became even more evident in interviewees' perceptions of how useful food policy councils are in addressing their objectives, which is explored later in the chapter.

SUCCESSES

Our study found that some of the ways food policy councils have been most effective is by supporting ongoing initiatives and by providing guidance and building capacity within new initiatives.

Accelerating existing projects

Several of the interviewees involved in preexisting initiatives noted that their food policy council accelerated their work through in-kind support, primarily staff time, and/or by facilitating strategic new connections between organizations or sectors. For example, one interviewee was engaged in a healthy corner store conversion initiative. Its aim was to increase the availability of healthy foods in corner stores in a neighborhood with few other retail food options. She explained how the food policy council was able to bring new expertise to this ongoing effort and serve as a neutral intermediary between good-food activists and corner store operators.

> I think that the Neighborhood Market Training would not have happened, at least not as soon and not as broadly without the food policy council.... The corner store operators, whether they have convenience stores, small groceries, or liquor stores sometimes have tense relationships with the communities that surround them.... So the food policy council, by offering this training, was a neutral entity bringing [corner store operators] into contact with people who design very successful niche grocery stores—people who are up to the minute on how you finance these things with small business lending, and could talk to them about sourcing locally or businesses strategies.... You could just see the light bulb going on in the room. It was great, great discussion. And there was a clear benefit for [the corner store operators]—they were getting access to this great information ... so the decision to show up was literally a business decision.

By leveraging relationships with both marketing and merchandising experts and community food security activists, the staff of one food policy council was able to create a neutral space, offer unique technical expertise to corner store operators free of cost, and help change the at times adversarial dynamics between proponents of the corner store conversion program and the corner store operators.

Similarly, another interviewee remarked that the food policy council had been very helpful in accelerating efforts to get farmers market operators to install Electronic Benefit Transfer (EBT) machines in order to take Supplemental Nutrition Assistance Program (SNAP) benefits (formerly known as food stamps) at farmers markets. Specifically, the council helped them engage farmers market operators: "They've co-convened several meetings with us where we brought together market operators. I think there are well over a hundred in the county and there's not a well-defined umbrella organization for them. [The council is] able to bring different folks together." In another example, a food policy council working group played an instrumental role in drafting the initial language for chicken and beekeeping ordinances. In doing so, they reduced some of the onus on the Legislative Reference Bureau and other city departments and were able to accelerate the passage of these ordinances.

These examples illustrate the potential of food policy councils to serve a facilitative role between different groups of stakeholders and between community groups and policy-making bodies. However, they also exemplify instances where partner organizations were, by and large, amenable to collaboration. McClintock and Simpson (chapter 5) highlight several instances in which departmental infighting thwarted food systems policy and planning efforts. Given that most food policy councils have little or no formal authority, further research is needed to examine the formal and informal recourse mechanisms that food policy councils have used in the face of interjurisdictional conflict.

Incubating and vetting nascent projects

Food policy councils have also provided valuable direction for nascent projects by vetting and strengthening initiatives originating from working groups to develop policies on topics such as street food vending and urban agriculture. Staff at one food policy council urged a working group to present its research and proposal to promote food carts before the appointed council. In this way, food policy councils can use their members' expertise to anticipate potential policy and regulatory barriers and leverage their professional networks to identify partners and strategies to help overcome them. Following is one of the working-group leaders' reflections on the experience of presenting a policy proposal to the council:

> They were totally in favor of it. They were really excited about it. And then they were like, "Maybe you should think about this," or "You should check out this report," or "If you need help in this part of the city, let me know." And so that was super valuable and actually got the working group really excited, like, "Wow, we have all of these other folks really supporting us, we can keep moving forward."

An appointed member of a council with a background in public health provided another example of this vetting function:

> [One of the working groups] presented to the council and I raised a concern about food safety—that a lot of this is unregulated. The environmental health folks are very rigid on this stuff. They come at this from a regulatory perspective. But the director of Environmental Health at the County Public Health Department is very open-minded and I said, "I think it would be good if you met with him . . . because I don't want to see the Health Department be an impediment. I mean, we certainly have a responsibility to ensure food safety, but I think there are ways of doing that while at the same time being supportive of the Urban Agriculture Movement." So they did meet, and I was told it was a very good meeting.

Fostering systems knowledge and partnerships

Another common theme that emerged is best captured by one interviewee who, in reflecting on what she had learned through her involvement with the food policy council, remarked, "I didn't even know what I didn't know." Many members come to their food policy council with knowledge or passion for a particular issue, such as urban agriculture or emergency food security, but relatively few members have previous in-depth knowledge of multiple food system issues. Further, few of the individuals interviewed in this study had previously served on committees with such a wide a range of goals and expertise. While this breadth of scope presents challenges for communication and agenda setting, it also fosters invaluable intangibles such as systems knowledge and partnerships between unlikely bedfellows.

For example, one council member who served as food service director for a large urban school district felt that school food often got a bad rap from people who did not fully understand the scale, price points, and regulations that structured school districts' sourcing and meal planning. She joined the food policy council partially out of a desire to educate others about the efforts the school district had made to trace and document its locally sourced products in spite of the scale and cost constraints it faced. As she explained,

> I think sometimes people are critical of school meal programs. So the food policy council provided me with a venue to educate people that might be local farmers or local producers—to give them an idea of the scope of my operation. But I think the learning went both ways, because it was also an opportunity to talk about what their capacity was, what our needs were, and to try and find some points of intersection or points that we could leverage.

While she joined the food policy council to clarify misperceptions about school food service, she encountered new perspectives and opportunities for cross-sector problem solving.

Food policy councils also foster knowledge exchange and partnerships by attracting members with complementary skill sets, for example technical assistance and lobbying. For instance, one interviewee noted that her professional work prevented her from directly engaging in lobbying. She found synergy with working group members who were able to "roll up their sleeves and visit with city departments and city council members." As she explained,

> I see where the needs are with urban agriculture, and I will develop educational materials and resources that support [the working group's] efforts. For example . . . how to remediate urban soils and what can be done around that. And it was clear that that's more of an educational role—teaching people about best practices and so forth. And I think it was going through that process of learning about what the city currently

did in regards to soil testing for example, and studying up on what was happening in other cities that really helped me see "Oh, this is an area where I could really focus some time because there's no one else who can do it."

In this way, she brought technical expertise through her knowledge of soil science and testing, while other members of the working group provided political voice to promote the expansion of urban agriculture. This is precisely the integration of policy advocacy and production expertise that Will Allen calls for in the foreword to this book as he reflects on what types of knowledge and collaboration will be necessary to overcome the myriad challenges facing urban agriculture; it is also consistent with Cohen and Wijsman's discussion of the coevolution of policy, planning, and urban agriculture in chapter 13.

Another interviewee reflected on how much he has learned about other facets of the food system through his involvement with the food policy council. He even pointed to ways in which his broadened understanding of the food system has potential implications for how public health interventions are conceptualized and approached:

> Our participation in the council has helped us expand the reach of our work . . . through introducing us to a broader network of potential allies. I have a medical background—a pretty traditional health and public health background. And in retrospect, I was remarkably uninformed about the food system and about the origin of food from farm to plate. . . . Through the food policy council, I've met a whole new group of professionals with different expertise, and it has helped us. I think we have a better understanding of what the issues are around getting healthy, affordable food to people. . . . So now I think we're much better informed when we're thinking about interventions. We need that broader understanding to engage these various players like small market owners, farmers markets, even schools.

Just as food policy councils have broadened the perspectives of their members, they could be used more deliberately to debrief local policy makers and municipal and county staff about innovations in food system policy and planning. As discussed in chapter 4, Horst, Brinkley, and Martin's research indicates that urban-agriculture policy is often spearheaded by urban growers and community members rather than municipalities. Citing the importance of governmental support for these change efforts, they suggest that there is a need for more "education and training for policy makers, planners, and public health officials" with regard to model codes and policies that could help support urban agriculture and community food security.

Finally, interviewees noted that participating in food policy councils has affected them personally. One council member mentioned that since joining the food policy council, she has started two home composting bins and built a greenhouse in her backyard for vegetable production. In reflecting on these projects, she remarked, "When you look at truly making change, you move from talking about it or thinking about it to doing it." This remark speaks to the fact that food policy councils have not only affected organizational activities, as demonstrated by the previous interview excerpt; they have also broadened individual participants' understanding of the food system and influenced the way they interact with it.

CHALLENGES

As the previous examples illustrate, at their best, food policy councils can harness the enthusiasm and creativity of newcomers to food system issues while leveraging the expertise and networks of experienced activists and mid- and late-career professionals to influence policy, enhance coordination, and build leadership. However, not all interviews pointed to such success. Criticism of council structures and processes tended to focus on three issues—how food policy councils interacted with existing organizations and activities, how meeting time was spent, and the underrepresentation of people of color and low-income individuals on food policy councils. Issues regarding the class and racial dimensions of food policy councils warrant their own discussion and are documented in a separate article. Here, the focus is on the first two critiques.

Most interviewees were active in the local food system prior to the creation of a food policy council, and some of them were wary about how their food policy council interacted with preexisting activities. Several interviewees—all of whom worked in the nonprofit sector—expressed concerns about food policy councils competing with other local organizations for grant funding. Others saw food policy councils as obstructionist organizations that were created by local government to dilute or displace progressive grassroots agendas. One interviewee with a background in antihunger advocacy pointedly asked, "Why can't the food policy council be something that gets support for stuff people are already doing instead of this whole new group of people parachuting in, and because of their connection with the mayor, sort of co-opting our whole agenda? A useful entity would help find funding and resources for existing groups." Indeed, as the successes described above suggest, one of the ways food policy councils have been most effective is by supporting ongoing initiatives and by providing guidance and building capacity within new community-driven initiatives.

In addition, some interviewees expressed frustration about how meeting time was used and complained about a lack of results, which they attributed in part to the variation in council participants' relevant experience and decision-making power. This frustration was especially apparent in the food policy council with an open membership policy. As one seasoned member explained, "On one end of the spectrum you've got mature organizations that are very clear on what they want to do and how they're going to do it. Then there are also a lot of folks in the room who just have this vague notion that they want to improve food."

Interviewees identified two main challenges with the open membership structure, as illustrated in the quotes below. First, bringing newcomers up to speed takes valuable time out of each meeting, which is not insignificant if the council meets only bimonthly or quarterly. Second, in order to have influence, food policy councils either need formal authority to make recommendations to local government or they must be made up of decision makers who can influence resources and practices at their respective organizations. Food policy council members with extensive professional experience in community food systems development and service provision were vocal about their frustration with the lack of focus and strategy that resulted from an open structure.

> It's an anybody can come anytime kind of meeting. And the difficulty that I have with that is how many times we go back to square one at a meeting. One of my great frustrations with the council is that it's essentially a networking meeting. I want to move something. I'm looking for

access points and partners to help push particular policy issues. . . . I really want this council to grow into a more formalized body that works to push on policy.

This quote underscores the frustration experienced by individuals who felt that a permeable, big-tent council structure impeded strategic policy work. The following observation from another interviewee raises questions about the efficacy of a food policy council made up of individuals with little formal decision-making power:

I'm not discounting the value of having the grand assembly. . . . But I think it's a mistake to think that the group could really affect policy without having a meeting of either board members or executive directors of the agencies involved in the work. The people who show up at meetings often can't really represent their organizations. Even if they believe strongly in something, it won't necessarily result in action when they get back to the office.

Relatedly, several experienced council members made it clear that while they value the space created by the food policy council, they already have high-level contacts and are much less dependent on the council to serve as an information clearinghouse or networking forum than newcomers. One interviewee commented, "I've been doing food work for so long that if I think I need to know somebody, I just pick up the phone and call them."

Multiple interviewees were also active in other coalitions or networks whose narrower geographic and/or issue focus seemed to foster greater trust or at least more expedient results. As another explained, "There's really just a handful of us working on [this topic] and so we're meeting and talking all the time anyway. We met before the food policy council existed and we still network on our own much more frequently—so the food policy council, for us, ended up being a little duplicative."

DISCUSSION AND RECOMMENDATIONS

This exploratory study suggests that people are motivated to participate in food policy councils to increase coordination with other organizations active in the local food system, to advance specific goals or projects, and to network and acquire information about existing activities. Not surprisingly, our findings indicate that individuals with more experience in local food issues tend to be motivated by a desire to advance specific initiatives and goals and seek strategic partnerships in order to influence policy and bring about systems change. By contrast, newcomers to community and regional food systems and individuals with fewer professional connections are more likely to seek general information and networking opportunities through their involvement with food policy councils. Further, as some of the more seasoned council members indicated, failure to meaningfully engage them and advance their goals may result in the declining involvement of food policy council members whose knowledge and networks are important for the development of emerging leaders and the efficacy of food policy councils as a whole.

As such, food policy councils must honestly assess whether their structures and processes align with their goals and those of their members. If the primary goal is networking, an open membership structure and informal relationship with local government may be perfectly appropriate. However, to significantly influence policy and systems change, food policy councils will likely require formal authority as recognized by local or state government, and they may also need to convene individuals who have decision-making power and influence within their respective organizations. Food policy councils that do not meet these criteria can serve as valuable information clearinghouses and networking organizations, but according to our preliminary findings, they may be more attractive to newcomers than to individuals with more established professional networks.

In spite of these challenges, our research shows that food policy councils have already added value to emergent and ongoing community food initiatives and have successfully supported both new and seasoned food activists and professionals through accelerating existing projects, incubating and vetting nascent projects, and fostering systems knowledge and partnerships. It is notable that while these activities are likely augmented by informal networking, what they prioritize is the development of strategic relationships in the service of specific objectives. This is illustrated by the food policy council that brought together store owners and grocery marketing professionals through a corner market training. Other examples include introducing citizen working-group members to specific staff in the Legislative Reference Bureau or County Health Department to strengthen grassroots program and policy proposals. Focusing on ways food policy councils can strategically support preexisting activities not only benefits new and seasoned food activists and professionals but may also help dispel claims that food policy councils are competing with local nonprofits for funding and turf.

RECOMMENDATIONS

1. *Develop an appointment-based membership policy with open-membership working groups.* In this study, the food policy council with the open-membership policy was not more demographically diverse or representative than the council with appointed members. Reserving certain council appointments for representatives of specific topics (e.g., farm to institution) or specific population segments and neighborhoods can help ensure more diverse representation. Meanwhile, open-membership working groups can create pathways for nonappointed residents, activists, and professionals to influence programs and policies through specific projects supported and vetted by appointed council members. To maximize public accessibility, open-membership working group meetings could be held in the evenings.

2. *Promote inclusive leadership development within food policy councils.* The central aim of an open-membership structure is to promote inclusivity and diversity, yet interviews indicated that while open-membership structures led to greater numbers of newcomers, they failed to enhance the cultural and demographic diversity of the council membership. In addition to using council structure to facilitate inclusivity and diversity, food policy councils can promote inclusive leadership development programmatically by building pathways for youth engagement, holding evening meetings, and hosting trainings on dismantling racism and implicit bias for council members and staff (Day Farnsworth, forthcoming). In cities where councils are required to use Robert's Rules of Order, trainings on meeting protocol have also been used to help level the playing field.

3. *Provide brief but regular newcomer orientations and publicize them widely.* Depending on how frequently a council and its working groups meet, it may consider hosting orientations before each council meeting or prior to working group meetings to answer frequently asked questions and to foster informed and productive participation from the public.
4. *Develop a process by which food policy council working groups can fine-tune project proposals and solicit endorsement from the council.* One of the greatest strengths of most food policy councils is the diverse expertise of their members; this is an invaluable asset to initiatives whose implementation and regulation will fall under the purview of multiple city or county agencies. Interviews indicated that while some councils have served a useful vetting function, the process and expected outcomes were often unclear. Food policy councils should consider developing processes through which working groups can pursue council endorsement for policy proposals as well as criteria for the types of projects the council is able to endorse; this may not include all the projects incubated in working groups.
5. *Establish grants programs to support and elevate community-led food projects.* For example, the Food Policy Council of Madison, Wisconsin, operates a $50,000 grants program to support community-based projects that increase healthy food access. Funded projects have ranged from youth gardening programs to the installation of salad bars in public school cafeterias.

CONCLUSION

Decentralization and misalignment of food policy and programming at the state and local levels have resulted in confusion and contradictory food system incentives and activities. By convening diverse stakeholders to address concerns pertaining to a host of food-related issues, food policy councils can use their members' diverse knowledge to help improve regulatory alignment and problem solving across the different agencies and sectors that shape our food system. Drawing on exploratory research, we have discussed what motivates individuals to participate in food policy councils and how their goals are addressed (or not) through participation in food policy council activities.

Preliminary findings suggest that food policy councils can enhance ongoing community food system work and build on their members' expertise and enthusiasm by accelerating existing projects, incubating and vetting nascent projects, and fostering systems knowledge and strategic partnerships. Further, we found that membership structure may influence a council's ability to effectively serve these functions. This is because individuals with more experience in local food issues tend to be motivated by a desire to advance specific goals, while individuals with fewer professional connections are more likely to seek general information and networking opportunities through their involvement in a food policy council. Although open-membership structures increase inclusivity, the councils that we examined found that open-membership structures resulted in a need to devote considerable meeting time to orienting newcomers rather than addressing substantive issues. Consequently, more-experienced members thought that the meetings were unproductive. To both nurture new food system leadership and harness the influence and expertise of experienced council members, food policy councils may wish to explore the following approaches: supplement appointed or elected membership structures with working groups that are open to the public, use programming to promote inclusive leadership development, host regular newcomer orientations prior to council meetings, develop a process by which proposed initiatives and policies can be formally vetted and endorsed by food policy councils, and establish grants programs to provide seed funding for grassroots food projects.

NOTES

1. Growing Power is a nationally renowned urban-agriculture nonprofit headquartered in Milwaukee, Wisconsin. It works to provide equal access to healthy, high-quality, safe, and affordable food for people in all communities through hands-on training and demonstrations in sustainable food production, processing, marketing, and distribution. Its founder and CEO, Will Allen, was the recipient of a MacArthur "Genius Grant" in 2008.

ACKNOWLEDGMENTS

A huge thank-you to USDA NIFA Agriculture and Food Research Initiative Award 2011-68004-30044 and the Evaluating Innovation and Promoting Success in Community and Regional Food Systems Project for support of this research. Thanks also to the many busy food policy council members and staff who took time to share their experiences, insights, and contacts with me. Last but not least, thanks to the reviewers in the University of Wisconsin–Madison Urban and Regional Planning Markets and Food Systems course in fall 2013 and 2014 for your feedback on earlier drafts of this chapter.

DISCUSSION QUESTIONS

1. Competition for resources and power can and has led to issues among organizations that participate in food policy councils. To what extent—if at all—is competition helpful or necessary in spurring community food system development?
2. Should a food policy council simply act as a facilitative chamber for its members, or should it initiate projects? Or does it depend? Explain.
3. Is open membership necessary for a food policy council to be fully inclusive? What strategies can food policy councils use to counteract problems that Day Farnsworth mentions as a result of a food policy council's open-door policy?
4. Is there a discernible hierarchy among food policy councils regarding organizational structure and sophistication? Should there be? What advantages and disadvantages are associated with different food policy council structures? Explain.
5. One purpose of food policy councils is to convene a wide variety of leaders with diverse expertise. However, such diversity sometimes leads to tensions. In what ways is this diversity helpful or detrimental to the work of food policy councils? What strategies does Day Farnsworth propose for navigating this tension? What other approaches could be employed?
5. How should a food policy council position itself relative to other

community food organizations, knowing that important community figures may believe that the food policy council's mission is redundant?
6. Given Day Farnsworth's chapter, consider this question: As organizations, are food policy councils too general, or too specific?

CHAPTER 16

Embedding Food Systems into the Built Environment

JANINE DE LA SALLE, PRINCIPAL, URBAN FOOD STRATEGIES

INTRODUCTION

The increasing sophistication and complexity of the global food system presents a paradox: on one hand, the planet has never produced as much food as it does now, and on the other, the rates of hunger, malnutrition, obesity, diabetes, and heart disease have soared in the Global North over the past three decades (Food and Agriculture Organization 2012). These health impacts are concentrated in urban environments. For planners, health practitioners, and other city builders, this paradox creates a strong impetus to not only understand the inextricable links between food, health, and the urban built environment but also to integrate regional food systems into the built environment as a core population health strategy. As a result of the significant health, economic, and environmental benefits linked to strong regional food systems, food provides an important lens or key organizing principle for the healthy built environment.

Prerail cities were (are) fundamentally shaped around regional and international food systems. From using fields outside the city for fattening cattle, to establishing ports that imported and exported foods from distant lands, to designing streets and creating food districts, early urban planners, architects, and landscape architects responded to the regional and international food system in the planning and design of streets, buildings, and open spaces prior to technological developments in transportation systems. In this way, prerail cities have been shaped by regional and international food systems from their earliest days (Steel 2008). Modern agriculture and global food systems have also facilitated a population shift to urban areas by enabling food to travel longer distances, be stored for longer periods, and ultimately feed a growing number of people.

As food systems globalized, transportation and storage technologies evolved as well, and cities were no longer constrained by largely regional food systems. City-building professionals, including urban planners, engineers, architects, professors, and politicians, have since fallen silent on the connection between health, food, and the success of towns and cities; indeed, these professions have traditionally treated food much like sleep: it is necessary but not meant to be regulated or managed in any meaningful way.

Urban planners and health authorities are beginning to collaborate on identifying the key strategies, performance metrics, and evidence-based research relating to these linkages in order to support local government decision making and community development. Specifically, more attention is being paid to understanding the links between the urban built environment, health, and regional food systems, as part of developing preventive health strategies. While there is a large bibliography on the many dimensions of health and the built environment (e.g., housing, transportation infrastructure, buildings, etc.), including food as a link is a relatively new area of exploration (Tucs and Dempster 2007). This chapter explores the links between health and the built environment, draws from North American case studies, and provides a framework for developing local strategies to build a healthy environment.

LINKS BETWEEN HEALTH, FOOD, AND THE BUILT ENVIRONMENT

The links between health, regional food systems, and the built environment are manifested in behaviors that are shaped by the physical environment. For example, walking to the corner store or green grocer to buy nutritious food decreases the need to drive, improving air quality and increasing opportunities for physical activity and social interaction. Planning for more walkable and compact communities also reduces development pressure on farmland on the periphery of growing metropolitan areas. Table 12 describes a range of health outcomes as they relate to opportunities in the healthy built environment. Strategies to create a healthy built environment vis-à-vis food systems are discussed in the following sections.

A NEW LEXICON FOR PLANNING AND DESIGN

Food in the built environment has been understood largely in terms of geographic access to food sources including grocery stores, community gardens, and so forth. In many ways, drawing the linkages between health and the built environment is a return to the emergence of urban planning as a response to a public health crisis caused by unhealthy environments. Because of the spread of contagious diseases resulting from overcrowding in tenement buildings in New York City at the turn of the last century, new regulations for land use and buildings were imposed to control both private and public spaces. These changes gave birth to zoning, and the regulatory landscape for the built environment in the West was changed.

Health Canada developed a comprehensive definition of the built environment:

> The built environment is part of the overall ecosystem of our earth. It includes the land-use planning and policies that impact our communities in urban, rural, and suburban areas. It encompasses all buildings, spaces, and products that are created or modified by people. It includes our homes, schools, workplaces, parks/recreation areas, business areas and roads. It extends overhead in the form of electric transmission lines, underground in the form of waste disposal sites and subway trains, and across the country in the form of highways. (Health Canada 1997, in Tucs and Dempster 2007)

Because of the significant health, economic, and environmental improvements food directly influences, the food system is offered here as a key organizing principle for comprehensive planning and design efforts. A food

Table 12. Health outcomes

Health outcome	Healthy built environment opportunity	Strategy (discussed in text)
Protect farmland: Creating compact, higher-density communities decreases the development pressure on farmland and infrastructure costs for municipalities. Farmland and farmers provide fresh, healthy food to regional markets and often donate large quantities of food to the charitable sector.	Establishing urban containment boundaries, urban/rural edge guidelines, compact community policies, buffer guidelines, trail access on the edge of farmland, agricultural land preservation policies.	*Strategy 1: Integrate regional food systems in growth management policies and practices.*
Increase access to healthy affordable food: Physical health and social interaction are positively impacted by the ability to walk to food assets like the green grocer or farmers market. Community gardens and urban farms provide places to socialize and learn food and farming skills.	Developing policies and design guidelines to encourage green grocers, cafés, pubs, and restaurants to locate in highly visible pedestrian-oriented areas. Creating urban design standards to encourage food activities and establishment of food assets, such as community gardens and urban farms.	*Strategy 2: Increase geographic access to food assets.*
Increase the experience of food: The experience of food, from growing and eating, to education and waste recovery, in towns and cities draws people together around food activities. Urban greening with food plants also provides natural infrastructure to clean the air, water, and soil in urban areas. Green buildings, including extensive and intensive rooftops, increase energy performance and reduce emissions.	Adopting green building guidelines and rating systems that include food spaces, and creating spaces for food celebration and discovery in the public and private realms. This may include requirements for wide sidewalks for grocers and cafés to "spill out" onto; policies to allow patios, sandwich boards, and awnings; and implementation plans for rooftop gardens, among others.	*Strategy 3: Design food systems into buildings, open spaces, and streets.*

Integrate farms with communities: Living in a community where fresh food is grown on-site and delivered to residents is increasingly an appealing lifestyle for home buyers and renters. Including other food system elements such as food processing and storage, and waste recovery facilities creates further opportunities for innovation.	Allocating up to 20 percent of a site for urban development, with the remaining areas reserved for agriculture in perpetuity. Integrating food system elements into the overall design, including growing spaces, rurally inspired architecture, market and processing areas, recreation trails around farm sites, and narrow lanes. Creating a development model that pays for the farmland and capital investments required to make the farm viable.	*Strategy 4: Develop and adopt rural design principles for "agrihoods."*
Provide recreation, rehabilitation, therapy, and community development opportunities: Gardening is one of North America's leading pastimes, and providing gardening opportunities allows residents of all ages and backgrounds to exercise, physically or mentally rehabilitate, and create relationships with other gardeners. Urban food gardens also improve ecological health by providing habitat and food sources for birds and pollinators.	Designing space for front and/or backyard gardens, including raised beds for low-mobility people, community gardens and orchards, linear fruit/nut orchards, edible forests, urban farms, and demonstration gardens, among others.	*Strategy 5: Expand urban food production activities.*
Stimulate an inclusive sense of place and community: Planning and designing multiple food assets into the built environment creates a critical mass of activities that gives unique identities to neighborhoods, cities, and regions. Creating places where people of different backgrounds and ages can congregate creates community and breaks down barriers of social isolation.	Designing public squares, plazas, and fairgrounds where food events such as agricultural fairs, farmers markets, and food truck festivals can be held.	*Strategy 6: Increase community identity and sense of place.*

Figure 18. Food system elements.

system may be described as the interconnected stages that food goes through from primary production to waste recovery. Food systems exist on multiple scales and within local, regional, provincial, national, and global boundaries. Figure 18 illustrates six core food system elements.

Within the food system, food assets are specific aspects of the built environment including resources, facilities, services, or spaces that are linked to creating healthy communities. Examples of food assets in the built environment include the following: food retail (grocery stores), green grocers, farm stands, farmers markets, restaurants, caterers, pubs, and cafes; food trucks and mobile food markets; community gardens and edible landscaping; urban farms and rural farms; emergency food programs and social services; food education programs; community kitchens; plazas and open spaces for food celebration; agritourism locations; food-processing facilities (nonprofit and commercial); food recovery infrastructure; food distribution facilities; food hubs and districts; and wet markets and fisherman's wharfs.

CHALLENGES AND BENEFITS OF INTEGRATING FOOD SYSTEMS INTO THE BUILT ENVIRONMENT

Challenges for increasing health through integrating sustainable food systems into the built environment are systemic. Before any significant and enduring changes can be introduced in the physical environment, a policy foundation must be established. The lack of policies that enable the establishment and success of the food systems is a key barrier.

A second challenge is the food system's complexity in terms of roles and responsibilities for implementing intersectoral, intradepartmental initiatives. Many strategies discussed below require broad collaboration between all levels of government, universities, developers, and community groups to achieve greater benefits.

A third, and more difficult, challenge is the lack of data and evidence that illuminate the causal factors between the built environment, health, and food systems. While the health impacts of activities like urban agriculture are increasingly being studied, as presented in chapter 14 in this volume, generally this empirical evidence base is still developing.

Despite the lack of data linking food system interventions in the built environment with positive health outcomes, the probable associations between food, health, and the built environment are undeniable. Air quality, movement patterns, access to amenities, urban green spaces, and walkability are all factors in urban environments that directly affect individual and community health. As the causal links between food systems, health, and the built environment are highly complex and multivariate, they "can be more readily likened to a tangled fishing net than a solid set of links in the chain holding the anchor" (Tucs and Dempster 2007). In this way, practitioners are urged not to get caught in the net of conventional evidence but to use probable associations to develop responses to preventing diet-related illnesses through the built environment. Research and censuses must continually be conducted to ensure a solid information base to inform sound decision making.

The benefits of including food systems in the urban built environment are wide ranging and, in short, include the following:

- Improving population health indicators (e.g., obesity) by developing walkable, active, transportation-based communities and increasing access to healthy food.
- Increasing access to fresh food sources (e.g., green grocers, community gardens, food hubs, farmers markets).
- Increasing capacity for local foods to be grown/raised/made and distributed.
- Creating a stronger connection between food and people, who will in turn be more engaged and informed consumers and appreciative of a healthy, multifunctional rural landscape (Francis et al. 2005).
- Creating community and magnetism around food places.
- Reducing food waste.

In response to these challenges and benefits, governments, universities, developers, and community organizations are setting visions, developing policy frameworks, and implementing initiatives that help create food- and health-conscious cities. The goals, strategies, and case studies explored below provide a starting point for developing a framework to address the many health opportunities in the built environment.

FOOD-CONSCIOUS BUILT ENVIRONMENTS: GOALS AND STRATEGIES

If we work from "probable associations" between the built environment, health, and food systems, a diverse range of local and regional strategies can support the integration of healthy food systems into towns and cities. These strategies are based on five goals, or desired outcomes, for embedding food systems into the built environment.

1. Increase opportunities for people to grow more food in towns and cities.
2. Increase the number of food assets, including food sources, as well as food-processing and storage infrastructure.
3. Increase opportunities for people to experience, participate in, learn about, and share food.
4. Increase opportunities for people to celebrate regional foods and food producers.
5. Recover high-quality foods that would otherwise be wasted for those who need them most.

Local and regional governments, nonprofit organizations, universities, and developers have created and implemented a number of strategies that embed food systems and health strategies into the built environment. These strategies occur at different scales of planning and design, ranging from macrodiscussions around growth management and agricultural land protection to detailed building and site design.

Strategy 1: Consider food systems in growth-management policies

Determining the boundaries of urban development, especially in regions with agricultural land and assets, is the critical starting point for embedding food systems in the built environment. While containing urban development has been associated with increased housing prices (Jansen and Mills 2013), developing land-use policies to ensure an agricultural land supply and vibrant farm economy creates downstream opportunities for generating community health and wealth through food. Despite growth-management discussions often being polarized and containing bitter public debates over building versus preserving, many cities and regions (e.g., the city of Portland, Oregon, and the Capital Regional District, British Columbia) have developed policy frameworks to balance growth and agricultural land supplies.

Over thirty years ago, the city of Portland established an urban containment boundary that provided landowners with certainty around the zoning and land uses in both urban areas and adjacent farm areas. Inevitably, as Portland grew, this containment boundary was revised and expanded. However, the long-range plan to contain growth in existing areas allowed for a strategic approach to local farming, and now the Portland area has become one of the most innovative and vibrant regions in North America.

The Capital Regional District on Vancouver Island in southwestern British Columbia has adopted an urban containment boundary intended to preserve a significant amount of agricultural land in a rapidly growing metropolitan area. The boundaries have been challenged by developers but have largely provided an effective tool for containing growth and protecting land. The Capital Regional District and stakeholders are now collaborating on how to improve agricultural viability and manage the many challenges of having agriculture near growing towns and cities.

Strategy 2: Increase geographic access to food

One key challenge for healthy eating is the ability to access healthy food and healthy diet education. The links between low income and food insecurity are well documented (BC Ministry of Health 2011). For example, seniors may have equity in their homes but very little disposable income to spend on food. Similarly, many low-income areas do not have easy physical access to healthy food sources like green grocers or farmers markets, forcing residents to travel long distances. Encouraging and allowing a range of food assets through policies and design in communities allows for more opportunities to access healthy foods.

Another food access strategy that is taking North America by storm is the mobile fresh market. With over forty mobile markets in the United States and more in Canada, enterprising nonprofits are using donated or retrofitted buses and vans to bring fresh produce and other grocery items into neighborhoods that do not have grocery stores or food programs. While mobile markets are not technically part of the permanent built environment, they nonetheless respond to urban development patterns that restrict access to community amenities such as food stores and other food assets.

Strategy 3: Design food systems into buildings, open spaces, and streets

Buildings, streets, and open spaces make up the built environment of a town or city. Through developing incentives, policies, guidelines, and requirements to integrate food systems into the built environment, the places and spaces in which we live, work, and play can improve our health.

For example, as part of the Cascadia ecotopia concept discussed by McClintock and Simpson in chapter 5, the city of Vancouver recently adopted a new rezoning policy that requires large projects (over two acres) to develop a food system plan, among seven other requirements ranging from renewable energy to active transportation (City of Vancouver 2013). Essentially, the city requires that a minimum of three out of six suggested food assets be included in the food system plan that must be submitted as part of a rezoning application. These include the following: community gardens/orchards; edible landscaping; community kitchens; community food markets; food composting facilities; and facilities to support neighborhood food network activities.

This is a significant shift forward from the previous policy that contained guidelines for encouraging urban agriculture, and the response from developers has been strong, with some applications designing food assets into the project. The city of Vancouver's rezoning policy requiring a food systems strategy for large projects is now one of the most forward-looking policies in Canada.

Light industrial buildings are also needed to support regional food systems in terms of providing storage and distribution infrastructure for local products. For example, food hubs are a concept gaining significant popularity in North America, and they can be located in light industrial areas as well as in more commercial and pedestrian-oriented locations. A food hub is a place that brings together a wide spectrum of land uses, design strategies, and programs focused on food to increase access, visibility, and the experience of food within the city. The USDA defines a food hub as "a centrally located facility with a business management structure facilitating the aggregation, storage, processing, distribution, and/or marketing of locally/regionally produced food products" (USDA 2010).

Food hubs are designed to meet specific needs in a given region or municipality, and they differ from place to place (e.g., from an urban food experience/access/education hub to a more rural agrifood distribution hub). In the United States alone, 168 food hubs now exist as defined by the National Food Hub Collaboration. The central theme in all food hubs, however, is that many functions are located together to create opportunities and synergies around local food connections, a strong farm sector, and healthy communities. The resurgence of food hubs in North America is driven by a lack of food education services and distribution infrastructure that allow people to develop food skills, as well as growers and processors to increase direct marketing and processing opportunities.

EXAMPLES OF FOOD HUBS

The Stop: Community Food Centre (www.thestop.org) is located in Davenport West, one of Toronto's lowest-income neighborhoods. It started out in the mid-1970s as a food bank, providing emergency food relief to the community. Since then, The Stop has evolved into a community food center that complements its emergency food services with a range of capacity, skill-building, and educational programs centered on food. The Stop now sees hundreds of people shop at the Green Barn Market each week; almost two hundred community kitchens are hosted; over seven hundred children participate in a food-systems education program covering poverty, cooking, food security, and farming issues; the Green Barn Farmers Market has supported fifty farm vendors selling approximately $1.25 million in local produce; and annually, The Stop purchases approximately $30,000 worth of local food and $40,000 worth (22,000 pounds) of local, organic food (The Stop 2014).

Red Tomato (www.redtomato.org) is a food hub based near Boston, Massachusetts. After establishing as a distributor for one farm, the business grew into a marketing, product development, and brokerage organization for multiple farms. The Red Tomato hub facility is supplied by twenty to twenty-five or so small to medium farms. Currently, its primary customers are food retailers. A unique aspect of Red Tomato is the marketing and branding services offered to suppliers. Products are sold under the Red Tomato brand, but farms are also identified on the label.

Eat Oregon First (www.eatoregonfirst.com) is a producer-owned and -operated distribution hub on the outskirts of Portland, Oregon. Farmers bring products directly to the hub, where products are then aggregated in the 12,000-square-foot warehouse located on commercially leased property. The hub stores, portions, and prepares products before distributing them to 120 food-service clients. The hub also undertakes the product marketing.

In addition to food hubs, municipalities may set design guidelines and policies for open spaces and streets that integrate food systems. For example, a great "food street" may include wide sidewalks to allow patios, plazas for farmers markets and food celebrations, allowances for awnings to cover produce-market sidewalk sales, and designated places for food carts to operate. This allows food culture to grow and develop over time, with specific areas known for certain types of food or food activity. For example, designing streets to accommodate weekly farmers markets enables the market to provide a place where people meet farmers and learn about local farming. Markets have been linked to increased food access, quality of life, and increased local economic activity. To maximize these benefits, local governments can use zoning as a tool to expand markets (Morales and Kettles 2009).

Similarly, parks may be designed and programmed to include community gardens and orchards as well as edible landscaping. These spaces allow people to not only grow food but also interact, breaking down barriers of social isolation experienced by many seniors. The city of Seattle and Seattle Public Utilities recently provided land and start-up costs to establish the Beacon Food Forest. This seven-acre edible park is an initiative of a nonprofit group called Falling Fruit, which aims to sustainably grow and harvest produce from the park's thirty-five or so trees as well as provide community education on edible landscapes.

Strategy 4: Develop and adopt rural design principles for agrihoods

Not all communities consider themselves "urban." In fact, many people continue to live in the peri-urban or rural areas. Often these areas are forgotten when it comes to planning for form and character, and setting out rural design guidelines intended to integrate agricultural neighborhoods with surrounding farm areas is a subject of untapped potential. For example, setting out guidelines to maintain the rural character can support food systems by increasing the contiguousness of farm areas, minimizing building footprints, and providing food assets like dedicated areas for farmers markets or food-processing facilities. There are twelve agrihood planning and site design principles to consider:

- **Protect prime soils** and locate buildings on the areas with the lowest-quality soil. Prime soils are identified as class 1 to 3.
- **Plan with nature** and preserve the ecological value of the site or

region including wildlife corridors, sensitive habitat, tree stands, hedgerows, and ponds.
- **Limit and restrict nonagricultural activities** (e.g., industrial, residential, and commercial development). Proponents wishing to rezone are required, in addition to existing requirements, to provide a strategy that demonstrates how there would be a net gain to food production and farming. This would include the provision of and mechanisms for resourcing predevelopment, implementation, and operating costs.
- **Site buildings** so that they do not interfere with primary production, circulation, and access, or the capacity of the land to be productive.
- **Prioritize land uses** that directly support regional producers and processors.
- **Consider nonagricultural uses (e.g., industrial, residential, commercial)** only when all other options have been exhausted.
- **Minimize building footprint.** Food processing facilities should be limited in terms of footprint and location.
- **Prioritize on-farm processing** and require that a minimum of 50 percent of the product processed be from the farm (i.e., not a Twinkie packaging plant). Types of processing that should be considered include meat processing and butchering; washing, grading, and bagging facilities; dehydrating; packaging and distribution; and cold and frozen storage.
- **Anticipate and mitigate impacts on neighbors** by reinforcing good-neighbor policies (e.g., mitigating increased traffic, noise, dust, and smell).
- **Increase direct purchasing opportunities** for regional foods (e.g., farm-gate retail, small- to medium-scale farm stores, and wholesale distribution).
- **Allow signage** for farms to promote their business.
- **Allow small-scale cafés, restaurants, and pubs** on-site.

Strategy 5: Expand urban food production activities

As one of the most visible aspects of the local food movement, urban food production for both personal and commercial purposes continues to gain popularity in North America. Backyard and front yard gardens, community gardens, and edible landscaping all provide visibility, interest, and activity and require planning and design for long-term success. The links between urban food production, health, and the built environment are many. Gardening provides an important form of mental, physical, and even spiritual therapy. It also fosters social connections and a sense of place through creating an environment where people interact, share information, and hold small community events.

The emergence of urban farming in North America has expanded the creative and innovative ways people are using space and developing new business models. In Chicago, land that was once part of a thriving urban center has gone through a process of degradation to restoration. Farming in these areas is an ideal strategy with low land prices, ability to consolidate many parcels, and proximity to some infrastructure.

In other, denser cities, urban farming has found its way onto rooftops and into peri-urban areas, borrowed backyards, vacant lots, and interstitial spaces. Small-scale social enterprises and businesses are forming around urban farming. While urban farming and food production will never replace the need for rural agriculture and farmers, it provides an important learning ground for new farmers, an asset for entrepreneurs, recreational and social opportunities, and a direct link to healthy food. In order for these benefits to be realized, physical space is needed. Space opportunities in the built environment for expanding urban food production include backyard and front yard gardens; community gardens; orchards in parks and school gardens; intensive rooftop gardens and vertical farming; edible landscaping around buildings and in parks and boulevards; and urban farms on vacant land.

Facilities to support urban food production are important to consider and include tool and equipment storage; light processing and storage; and warehousing and aggregation areas. A straightforward yet important strategy for policy makers in enabling these facilities is simply to "get out of the way." For example, allowing the sale of produce from backyards removes a key barrier for urban farmers to efficiently connect to urban food buyers, thereby making production activities more viable.

Strategy 6: Increase community identity and sense of place

Food places and destinations create interest and activity, and they appeal to people across cultural, age, and gender groups. By creating magnetism or gravity in a specific area, food places create community cohesion and a sense of identity. On the waterfront of Vancouver, British Columbia, Granville Island is a food magnet where a vibrant wet and dry market, cafés, and restaurants draw hundreds of thousands of residents and visitors every year. Like other food magnets, Granville Island works because there is a diversity of food-related activities, businesses, and people. This has been facilitated by the intentional planning, designing, and programming of land, streets, buildings, and open spaces to make dynamic and highly energetic food places.

An applied example of a project that seeks to bridge health, food, and the built environment is the Pearson Dogwood redevelopment, also in Vancouver. A case study of this project, authored by the lead consultant planner, is presented below.

CASE STUDY: FOOD, HEALTH, AND SENSUALITY AT PEARSON DOGWOOD (JENNIFER FIX, DIALOG)

Rethinking health and health funding

Food, health, and sensuality intersect in one of the largest redevelopment projects in Vancouver, British Columbia. For the first time in Canadian history, a health authority is using the real estate value of its land assets to create new health services and amenities for its population. Through redevelopment of an underutilized but highly valuable 25-acre site called Pearson Dogwood, Vancouver Coastal Health is creating an endowment for much-needed capital projects across the region. In turn, the current residents—who live in two outdated institutions and require complex care because of disabilities ranging from multiple sclerosis to spinal cord injuries—will be provided with new housing and access to a range of health amenities directly on-site.

Once the site is developed, the new neighborhood is expected to include approximately five thousand residents living in both market and nonmarket housing, a YMCA, a community health center, a therapeutic pool, small-scale retail, parks and plazas, a new rapid-transit station, and more. The master plan for the new mixed-use community is driven by a vision for "whole health," which includes:

whole people in all facets of being, including physically, emotionally, mentally, and spiritually;

whole communities, including the diverse needs and aspirations of current and new residents, as well as the social connections that create a sense of community and belonging; and

whole ecologies, including a recognition that the health of individuals and communities hinges on the health of other species and the natural processes of which we are a part.

Food and whole health

One of the foundational building blocks of this "whole health" vision is food. As a significant health determinant, diet is one of the biggest factors in major noncommunicable diseases (World Health Organization 2014). The presence of healthy food—either grown or commercially available—in the future neighborhood is thus crucially important to its success as a place of health and well-being. Also, since nearly one-quarter of all trips in cities are associated with purchasing and consuming food, food destinations in the new neighborhood can also support active, healthy living by encouraging residents and neighbors to access food by foot, wheelchair, and bicycle.

The prevalence of active transportation also translates into ecological health benefits, with fewer greenhouse gas emissions than in neighborhoods that are more oriented toward vehicular transportation (Dodman 2009). At the same time, human health hinges on broader ecological health; for example, chronic exposure to traffic-related air pollution causes nearly nine times as many premature deaths in Canada as do traffic crashes (Brauer, Reynolds, and Hystad 2013). Likewise, healthy food—or at least food that is produced locally and organically—has a smaller carbon footprint and can contribute to soil health, habitat for other species, and overall healthier ecosystems.

In addition to being important to individual and ecological health, food at Pearson Dogwood is vital to the social health of the community. The site is currently very insular and has a somewhat stark, institutional character, with limited physical and social connections with the mature residential neighborhood that surrounds it. Using food as a tool to "turn the site inside out," the master plan envisions diverse community-oriented food destinations that draw people into the site. These destinations range from spaces for informal gathering and chance meetings (e.g., cafés and edible landscaping in parks and along paths) to more-formal programs and spaces for gathering (e.g., shared gardens). The latter builds on the precedent set by a small urban farm that currently exists on the site, consisting of a market garden, wheelchair-accessible community garden, and teaching garden that matches disabled residents with horticultural therapists and volunteer gardeners.

Food and sensuality

And still the links between food and health run deeper at Pearson Dogwood. During consultations with current residents about their desired future in the new neighborhood, people spoke about hearing birdsong, smelling aromatic flowers and herbs, feeling sunlight on their faces, and seeing people working in the vibrant colors of a food garden. Indeed, despite being bound to wheelchairs or beds because of complex disabilities, including paralysis, these residents called for a neighborhood experience that is intensely sensual.

The relationship between rich sensory experiences and health is important. Places that stimulate our senses are deemed to have health benefits that extend beyond our physiology. Evidence-based research tells us that there are strong links between mental and emotional health, and access to nature and green space (Thompson et al. 2011). According to Olga Ruskin, a resident of one of the institutions at Pearson Dogwood and a user of the urban farm (Rashleigh 2013), "This garden has changed my days. When I feel low, I can come out and look at the flowers and I feel happy because they are so pretty and colorful. They literally uplift me if I have a bad day. I can't do much because my hands are stiffened with arthritis, but I can go into the garden and get engrossed in the plants and forget my cares."

Food has an important role to play in the sensual experience of place, and in fostering health. Food is powerfully sensual, delighting us with color and texture, and emanating scents that can leave our mouths watering and evoke childhood memories. Encouraging a high degree of sensuality, the master plan's open space hierarchy includes a strong food and therapeutic garden component, with tactile, aromatic, and audible features that bring people into their bodies and celebrate them as feeling, conscious beings regardless of their physical abilities. The food system at Pearson Dogwood is designed to touch minds, spirits, and all types of bodies, engaging people's senses and helping them to feel, heal, and flourish.

Closing

The Pearson Dogwood project provides an inspiring and leading example for integrating food into a built environment that prioritizes health. Creating positive impact also occurs at a much smaller level and could be as simple as having tea with a colleague from across the hall to hatch an idea. As strategies for integrating food systems into towns and cities become more diversified and sophisticated, a wide range of health and community wealth benefits are possible. Reconnecting people with food in the built environment also results in establishing a new normal, propelling innovation and development of sustainable urban and regional food systems.

DISCUSSION QUESTIONS

1. How effective can city departments be in integrating farms in a major city proper? Is integrating farms necessarily related to place making? Explain.
2. Analyze the table in de la Salle's chapter. What's good about it? Is there anything missing? Are some proposals in this table more realistic than others?
3. The author states that policy must be in place before changing the physical environment. Is this true in all cases? When it is true, how does this put certain communities at a disadvantage? Which communities lose out?
4. How should cities and urban-agriculture actors adapt to a general lack of data when planning urban food systems?
5. Examine each of the five strategies that de la Salle presents for integrating food systems into cities and towns. What are the chief positives and negatives of each? Is there a strategy that would work for communities in general?

CONCLUSION

JULIE C. DAWSON AND ALFONSO MORALES

All the chapters in this volume relate urban agriculture to urban food systems and vice versa, and in doing so they illustrate the general principles cutting across food systems issues. Our purpose has been to provide concepts and examples of urban food production but also to show how its many activities are interwoven with the activities of distinct, often conflicting but possibly allied, organizations. Identifying opportunities for cooperation is important, but cooperation is impossible without taking the time to relate different ways of talking and thinking about those activities. What seems simple often is not, and it takes firm and flexible relationships to make identifying and achieving common goals possible.

Before going on to describe some important new questions for further research, we believe it essential to remember that not everything can be reduced to statistics or metrics. Case studies show in detail what is difficult to capture in survey methods. While economic value has always played a role in food system innovation, other variables such as human values and aspirations to build healthy, equitable, and thriving communities drive experimentation in urban agriculture. For us, the strength of the case studies presented here is to show in detail many of the values people hold in their activities, and how they attempt to put these values into action. The case studies also show

how those values come into conflict with regulatory frameworks and with different practices and values held by others. We hope this book provides examples that stimulate new ideas, discussion, and action toward building context sensitive partnerships among organizations that might otherwise see each other as competing. We look forward to seeing many more successful examples of urban agriculture in practice emerge.

SOME THOUGHTS ABOUT FUTURE RESEARCH

In keeping with the organization of the book, we offer here some thoughts about future directions for research on the subject of urban agriculture and how it connects with and relates to other elements of the food system. In doing so, we offer a few guiding principles for thinking about research on this topic. First, we must account for different perspectives among individuals practicing urban-agriculture activities. Demographic differences, along with experience, make for different points of entry and potentially different motivations and goals that should be considered. However, differing demographic backgrounds do not necessarily imply that people will divide along these lines. While it is not possible to represent absolutely every perspective in a volume such as this, we hope these examples have shown how people with very different backgrounds can unite around common goals or values.

Second, we think it important to contemplate different organizational perspectives, including those of government, community-based organizations, and profit-oriented organizations. Each of these organizations has its distinct interests, roles, and time horizons. Each directs its efforts toward relationships of different kinds. For us, again, understanding the dissimilarities is not as important as reasoning together about points of connection in these different perspectives. With respect to both individual and organizational differences, a researcher will take time to consider whether and how different participants in the activity of interest will be able to identify and understand similarities with other participants.

Third, those interested in urban agriculture must take care to understand regulatory conflicts and opportunities associated with different activities. Law, policy, and regulation all influence urban food production, processing, distribution, and consumption, and the transformation of waste into productive inputs. The perspectives held by mayors or other city officials, those of county or state officials, and those of other units of government may clash, and reconciling these distinct perspectives in order to produce law and policy generative of economic and social activity and supportive of public health is vitally important.

With these three guiding principles in mind, useful for interested scholars or community practitioners, we now turn to some suggestions for research we think would be useful to the field of urban agriculture.

History and community significance

How much do we know about the history of food production in cities? We have an understanding of how marketplaces were developed in conjunction with cities, and for many of the same purposes, large-scale food production is now returning to cities. However, as the historians in this book have identified, there is much to learn about the history of food production and how that history has influenced law and policy and urban-agriculture activities today. An understanding of the history, especially of how urban agriculture has coexisted with rural agriculture over time and how agriculture started to be seen as an activity that did not belong in the city, can help us understand how the issue is being reshaped to be accepted in different communities today, for example in suburban and superurban areas, low- and high-income neighborhoods, and small and large cities. We must remain aware, however, that the historical records may capture only a single perspective on urban agriculture and may not include many of the agricultural enterprises historically present in cities. In particular, we should look for the stories of those whose efforts may not be represented in the historical record, such as immigrant communities and communities of color. Knowing more about the history of urban agriculture will also improve our knowledge of how diverse communities engage with urban agriculture and how to increase the chances that urban-agriculture projects will build community within diverse neighborhoods. As Thrasher points out in chapter 10 in discussing the Brooklyn Community Garden, we have much to learn about cooperative governance and community relations surrounding urban-agriculture projects. In addition to a more detailed historical analysis of urban agriculture in cities in the United States, a comparison across countries would be useful to both practitioners and policy makers. Many countries have rich histories of urban agriculture, and we in the United States could learn from their experiences.

Regulation

Regulation connects politics and policy to the larger political economy. Over the last forty years the conflict between liberal and neoliberal governmental modes has played out, broadly speaking, with the outcome of increasing economic inequality and eroding services in communities. This book has not, and cannot, address this broader concern. In effect, urban food systems are in some ways a response to declining governmental support of society. How will communities and organizations work with government in this time of decreasing resources from government, and increasing competition for foundation and private funding? What shape will society take in terms of new social priorities and movement toward the local, and how will food systems help shape that new society? Such questions cannot be answered here, but regional sociopolitical organizations concerned with that future will help shape those answers, and our effort has been to help readers think about the contours of that future. We need bold experimentation that we can evaluate for its importance and influence on communities and society.

Urban-agriculture production

More research is needed on the particular advantages and disadvantages of farming in the city. Urban farms are proximal to people, and this can have benefits in terms of access to inputs, labor, and customers under the right circumstances. The potential for urban composting and water reclamation remains untapped to a large extent. Increasing urban composting activities for urban food production could address both soil-quality issues and urban waste management, yet it remains very difficult for urban farms to access or create high-quality compost from urban sources. A deeper understanding of the role urban agriculture can play in waste management in cities, for composting, water retention, runoff control, and productive use of vacant land or rooftops, would help policy makers ensure that regulations are not a barrier to urban agriculture while also protecting food and worker safety.

Water is an issue for many urban-agriculture sites because agricultural rates are hard to qualify for and residential rates may make farming unviable. However, the use of reclaimed or recycled water can support urban agriculture and reduce city water-treatment needs. For example, many cities in California are expanding their recycled water program. According to California state standards, tertiary recycled water can be used for all irrigation purposes. However, there have been issues with salinity and heavy metals in the water, which hinder production. Since reclaimed water might be the only way urban agriculture can remain viable in California, more research on the effects of reclaimed water on crop production and ways to remediate water so it can be used would be helpful. Most literature on this subject currently focuses on the use of reclaimed water in larger-scale production agriculture rather than the urban-agriculture context.

In addition to understanding the production needs of urban farmers in terms of inputs like compost and water, understanding the potential of urban food production is also important. In particular, there is a dichotomy between production of high-value products to ensure farm viability on small lots within urban areas, and the production of food with and for communities that lack access to fresh produce. Currently, these two activities tend to be lumped together in the general category of urban agriculture, but they may have very different production needs. The same organization may pursue both avenues, with higher-value products helping to increase community food security. But very few data exist about the actual quantity, production cost, and market value of urban produce. The value of urban agriculture should clearly not be measured only in terms of its productive potential or economic value. However, it would be very useful for practitioners to have a point of comparison for their own activities, and for urban planners to understand the productive capacity of urban land. Small-scale intensive agriculture can be extremely high yielding, but many studies of the food-production potential of home gardens or small-scale intensive plots are not done in urban soils or in industrialized cities. Benchmarking production would contribute a starting point for research on how urban agriculture could be scaled up to produce more food, perhaps connected to local institutions such as schools or neighborhood grocers.

Planning, policy and regulations, food policy councils, planning food systems

Part of the lack of data is because urban agriculture does not fall under a particular part of government, so there are no statistics or data collected on a regular basis. Following the recommendations in the chapter on food policy councils to give these bodies more governing capacity could include charging food policy councils (with funding for these activities) with collecting more information about urban-agriculture production and related impacts of urban agriculture. This could be accomplished in close collaboration with researchers, community-based organizations, other branches of government, and local businesses. In particular, it would be useful for policy makers and residents to know the impacts of urban agriculture on property values and the reduction of urban blight.

Identified throughout the volume are instances in which regulations obstruct the advancement of activities that support, promote, and are critical to urban-agriculture systems. Rectifying these barriers is necessary to supporting current practices and exploring innovative methods of production and distribution. Many of the benefits of urban agriculture relate to increasing the community presence on city streets, beautification, increasing land values, and rebuilding community knowledge networks, but these are more difficult to quantify than food production or the development value of urban lots. Cities have used urban agriculture and community gardens as a way to address problems with vacant lots, but there are few policies in place to protect urban agriculture from development pressure as neighborhood revitalization occurs. How can we differentiate situations where urban agriculture acts as a vacant land "placeholder" from those where it could be a broader community/economic development strategy?

In particular, research on land tenure arrangements for community gardens and urban-agriculture sites could help practitioners and policy makers identify best practices across communities. Lack of land tenure is one of the primary barriers to the establishment of gardens and urban farms, and while some research has been done on community garden tenure arrangements, there are few examples of land tenure arrangements for commercial urban farms. In addition to land tenure arrangements, a more diverse set of examples of best practices for cooperative governance of urban-agriculture sites or community gardens could help practitioners.

In a broader policy perspective, land-use decisions operate on a regional scale, and the relationship between rural agriculture and suburban or urban agriculture is poorly defined. Case studies such as the one on Regional Access in New York show that it is possible for a distribution company or food hub to operate on a regional level, with many small to medium-sized producers and buyers. This and other examples highlight the potential for increased connections between rural and urban agriculture. Such arrangements could help facilitate linkages between peri-urban farms and more-urban agriculture. Some nonprofits such as Growing Power in Milwaukee operate at sites within the city and on the urban fringe, choosing production based on the relative benefits of rural or urban areas (hoop houses on urban sites growing high-value produce, and peri-urban land used to produce carrots for Milwaukee schools, for example). We need to understand the potential for these linkages to increase regional self-sufficiency, and the relative benefits and costs of encouraging urban production of different crops.

Public health and urban agriculture

It is hard to demonstrate the direct impact of urban agriculture on public health outcomes because of the long-term nature of studies on human health and well-being and the multitude of factors that contribute to individual and population health. Proximity to growing food may encourage city residents to eat more healthfully, seek out farmers markets or other sources of fresh produce, join a community garden, or add a plot to their own yard or balcony. The educational and outreach potential of urban agriculture should not be discounted. Specific questions that may prove more tractable as we gain experience with urban agriculture include the following: How can urban agriculture be used to improve food access to underserved communities? What impacts does urban agriculture have on people's desire for fresh produce, and on the availability of fresh produce in informal markets? Often urban producers are motivated by a desire for high-quality food; does this individual motivation spill over to neighbors and community residents who might not have been initially concerned with access to fresh produce? Can urban agriculture promote linkages to other sources of fresh produce, such as farmers markets, or encourage local retailers to carry fresh produce for residents?

Finally, we must do a better job of documenting impacts. We have many examples of how food production enhances urban life by opening up green spaces, improving soil quality and water and waste management, providing opportunities for community building, making healthy, affordable food more widely accessible, and enabling people to grow foods culturally important to them. However, in the face of ongoing skepticism about the relevance of urban agriculture, we need more systematic efforts to document these benefits, as well as potentially negative outcomes, in order to provide policy makers and practitioners with recommendations for best practices for meeting different urban food system goals.

The health and livability of a city depend on a multitude of planning decisions and the actions of local businesses, community organizations, and individual citizens. The urban-agriculture case studies we have followed show how food production in the city intersects with many different aspects of urban economic development, community development, environmental sustainability, and human health. We believe this volume provides useful examples of ways that urban agriculture fits into the fabric of the modern city, and directions for future research and innovation.

REFERENCES

SECTION ONE: INTRODUCTION AND HISTORICAL ANTECEDENTS

Allen, P. 2004. *Together at the Table: Sustainability and Sustenance in the American Agrifood System*. University Park: Pennsylvania State University Press.

Brown, A. 2001. "Counting Farmers Markets." *Geographical Review* 91 (4): 655–74.

Castle, E. 1917. "Food Situation in Women's Hands." *Grand Rapids Herald*, July 8.

Chamberlin, E. 1918. Correspondence to the Board of Commissioners of Grand Rapids, Michigan. Grand Rapids Commission Proceedings, #2220, April 8.

Charity Organization Society. 1894. Minutes of the Charity Organizational Society, First Annual Meeting, June 8. Grand Rapids Public Library Archives, Collection 382, Series 1, Box 6.

Cronon, W. 1991. *Nature's Metropolis: Chicago and the Great West*. New York: W. W. Norton.

Cummins, S., and S. Macintyre. 2002. "A Systematic Study of an Urban Foodscape: The Price and Availability of Food in Greater Glasgow." *Urban Studies* 39:2115–30.

Danbom, D. B. 1979. *The Resisted Revolution: Urban America and the Industrialization of Agriculture, 1900–1930*. Ames: Iowa State University Press.

Davis, L. M. 2006. "The Road to Reform: Building Healthy and Safe Communities." PhD diss., University of Minnesota.

Delind, L., and A. Ferguson. 1999. "Is This a Women's Movement? The Relationship of Gender to Community-Supported Agriculture in Michigan."

Human Organization 58 (2): 190–200.

Deutsch, T. 2010. *Building a Housewife's Paradise: Gender, Politics, and American Grocery Stores in the Twentieth Century*. Chapel Hill: University of North Carolina Press.

Donofrio, G. A. 2007. "Feeding the City." *Gastronomica: The Journal of Food and Culture* 7 (4): 30–41.

Fitzgerald, T. J., and R. T. Ely. 1921. *Mortgages on Homes: A Report of the Results of the Inquiry as to Mortgage Debt on Homes Other Than Farm Homes at the Fourteenth Census, 1920*. US Department of Commerce, Bureau of the Census. Washington, DC: US Government Printing Office.

Forsyth, A., M. Hearst, M. J. Oakes, and K. H. Schmitz. 2008. "Design and Destinations: Factors Influencing Walking and Total Physical Activity." *Urban Studies* 45:1973–96.

Garvey, T. J. 1978. "The Duluth Homesteads: A Successful Experiment in Community Housing." *Minnesota History* 46:2–16.

Grand Rapids Board of Education. 1915. *Annual Report*. December 12, 319.

———. 1917. *Annual Report*. June 23, 154.

Grand Rapids Board of Trade. 1888. *Grand Rapids as It Is: Illustrated Publications*, 2nd ed. and *Grand Rapids: Illustrated Publications*. Grand Rapids Public Library Special Collections, Box 13, Folder 96; Box 14, Folder 100.

Grand Rapids Chamber of Commerce Miscellaneous Committees. 1915. *Report of Agricultural Department*. April 22. Grand Rapids Public Library Archives, Collection 46, Box 39.

Grand Rapids City Council Proceedings. 1906. #28405, April 16.

Grand Rapids Federation of Women's Clubs. 1916. December 13. "Minutes of Special Meeting." Grand Rapids Public Library Archives, Collection 020, Box 7, Folder 1.

Grand Rapids Ladies Literary Club Minutes. 1903–1904. March 5, April 2, May 14, October 3. Grand Rapids Public Library Special Collections, Box 275, Folder 13.75.

Grand Rapids Public Service Department. 1924. *Annual Review of Markets*. Grand Rapids, MI. May 8.

Hamilton, E. M. 1917. *Committee on Market Sites Report*. Correspondence to the High Cost of Living Commission. Grand Rapids Common Council Proceedings, #72615, May 19.

Haskins, F. J. 1912. "Work of the Women's Clubs, Part 2: In Civic Work." *Grand Rapids Press*, July 3.

Hollister, J. 1917. Letter to Caroline Bartlett-Crane, June 26. Western Michigan

University Archives, Caroline Bartlett-Crane Papers, A92, Box 13.

Howell, C. 1897. "Vacant Lot Farms." *Grand Rapids Democrat*, February 26.

Hudelson, R., and C. Ross. 2006. *By the Ore Docks: A Working People's History of Duluth*. Minneapolis: University of Minnesota Press.

Hurt, R. D. 2002. *American Agriculture: A Brief History*. Rev. ed. West Lafayette, IN: Purdue University Press.

Kleiman, J. D. 2006. *Strike! How the Furniture Strike of 1911 Changed Grand Rapids*. Grand Rapids, MI: Grand Rapids Historical Commission.

Koc, M., and International Development Research Centre. 1999. *For Hunger-Proof Cities: Sustainable Urban Food Systems*. Ottawa: International Development Research Centre.

Lawson, L. J. 2005. *City Bountiful: A Century of Community Gardening in America*. Berkeley: University of California Press.

Lee, G., and H. Lim. 2009. "A Spatial Statistical Approach to Identifying Areas with Poor Access to Grocery Foods in the City of Buffalo, New York." *Urban Studies* 46:1299–1315.

Leonard, C. H. 1911. Correspondence to the Common Council of the City of Grand Rapids, Michigan. Grand Rapids Common Council Proceedings, #50563, September 11.

Lindeman, E. C. 1916. *The Grand Rapids School Home Garden Plan for School Children*. Grand Rapids, MI: Grand Rapids Board of Education.

Lippincott, J. B. 1920. "City Dads Talk School Gardens." *Grand Rapids Press*, March 30.

Lyson, T. A. 2004. *Civic Agriculture: Reconnecting Farm, Food and Community*. Medford, MA: Tufts University Press

———. 2005. "Civic Agriculture and Community Problem Solving." *Culture and Agriculture* 27 (2): 92–98.

"The March of the Cities: Duluth and Its Hinterlands." 1911. *The World's Work: A History of Our Time* 23:118.

Mattocks, N. 1911. *Greysolon Farms Co.* Duluth, MN: O. F. Collier Press.

Melosi, M. V. 2008. *The Sanitary City: Environmental Services in Urban America from Colonial Times to the Present*. Pittsburgh: University of Pittsburgh Press.

Michigan Association for the Prevention and Relief of Tuberculosis. 1909. *Annual Report: 1908–1909*. February 26.

"A Model Co-operative Marketing Association." 1911. *The Farmer: A Journal of Agriculture* 17:613–17.

Morales, A. 2000. "Peddling Policy: Street Vending in Historical and Contemporary

Context." *International Journal of Sociology and Social Policy* 20 (3/4): 76–99.

Mougeot, L. J. A., and International Development Research Centre. 2006. *Growing Better Cities: Urban Agriculture for Sustainable Development.* Ottawa: International Development Research Centre.

Nordahl, D. 2009. *Public Produce: The New Urban Agriculture.* Washington, DC: Island Press.

O'Neil, D., C. Heitman, A. Galletti, et al. 2008. "Expanding the Fulton Street Farmers Market: A Feasibility and Strategic Business Plan." Project for Public Spaces, October 25.

Pack, C. L. 1919. *The War Garden Victorious.* Philadelphia: J. P. Lippincott.

Paez, A., R. G. Mercado, S. Farber, C. Morency, and M. Roorda. 2010. "Relative Accessibility Deprivation Indicators for Urban Settings: Definitions and Application to Food Deserts in Montreal." *Urban Studies* 47:1415–38.

Peters, J. 1911. Correspondence from Cabinet Makers Union No. 1369 to the Grand Rapids Common Council. Grand Rapids Common Council Proceedings, #50879, October 14.

Polk, R. L. 1893. *R. L. Polk and Company's Grand Rapids Guide.* Grand Rapids, MI.

Reganold, J. P., D. Jackson-Smith, S. S. Batie, R. R. Harwood, J. L. Kornegay, D. Bucks, C. B. Flora, et al. 2011. "Transforming U.S. Agriculture." *Science* 332:670–71.

Rynbrandt, L. J. 1997. "The 'Ladies of the Club' and Caroline Bartlett Crane: Affiliation and Alienation in Progressive Social Reform." *Gender and Society* 11 (2): 200–14.

Sparks, A. L., N. Bania, and L. Leete. 2010. "Comparative Approaches to Measuring Food Access in Urban Areas: The Case of Portland, Oregon." *Urban Studies* 48 (8): 1715–37.

Stark, S. L., D. Abazs, and D. Syring. 2011. *Defining the Agricultural Landscape of the Western Lake Superior Region: Realities and Potentials for a Healthy Local Food System for Healthy People.* Final report submitted to the Healthy Foods, Healthy Lives Institute, Department of Food Science and Nutrition, University of Minnesota, 1–50.

Stockbridge, F. P. 1913. "Two Cities That Turned Farmers." *The World's Work: A History of Our Time* 25:459–69.

Syring, D. 2012. "Exploring the Potential for a More Local Food System in the Western Lake Superior Region." *CURA Reporter,* Fall/Winter, 10–16.

Tangires, H. 2003. *Public Markets and Civic Culture in Nineteenth-Century America.* Baltimore: Johns Hopkins University Press.

Tarr, J. A. 1984. "The Evolution of Urban Infrastructure in the Nineteenth and Twentieth Centuries." In *Perspectives on Urban Infrastructure*, edited by R. Hanson, 4–66. Washington, DC: National Academy Press.

Thompson, M. J. 1938. *The First Twenty-Five Years of the Northeast Experiment Station, Duluth*, 1–12. Duluth: Institute of Agriculture, University of Minnesota.

———. 1954. *The First Forty Years at the Northeast Agricultural Experiment Station, 1913–1953*, 1–32. Duluth: Institute of Agriculture, University of Minnesota.

———. 1959. *85 Years of Farming in the Northern Coniferous Forest Areas of Minnesota, Wisconsin and Michigan*, 1–58. Duluth: Institute of Agriculture, University of Minnesota.

Tilma, G. P. 1917. Correspondence to the Common Council of the City of Grand Rapids, MI. Grand Rapids Common Council Proceedings, #726303, March 26.

US Census Bureau. 1913. *Thirteenth Census of the United States: 1910*. Vol. 4, *Reports by States for Counties, Cities, and Other Social Divisions*. Washington, DC: US Government Printing Office.

Van Brunt, W. 1921. *Duluth and St. Louis County, Minnesota; Their Story and People*. New York: American Historical Society.

Van Buren, F. 1913. "Grand Rapids School Gardens." *Nature-Study Review*, April 13, 95–100.

Van Cleef, E. 1912. "A Geographic Study of Duluth. Part II." *Bulletin of the American Geographical Society* 44 (7): 493–506.

Viljoen, A., and J. Howe. 2005. *Continuous Productive Urban Landscapes: Designing Urban Agriculture for Sustainable Cities*. Oxford: Architectural Press.

Vitiello, D., and C. Brinkley. 2013. "The Hidden History of Food System Planning." *Journal of Planning History* 13 (2): 91–112.

Watson, S. 2009. "The Magic of the Marketplace: Sociality in a Neglected Public Space." *Urban Studies* 46:1577–91.

Whelan, A., N. Wrigley, D. Warm, and E. Cannings. 2002. "Life in a 'Food Desert.'" *Urban Studies* 39:2083–100.

Wrigley, N. 2002. "'Food Deserts' in British Cities: Policy Context and Research Priorities." *Urban Studies* 39:2029–40.

Zuller, E. 1911. "Push the Market Along." *Grand Rapids Press*, May 11.

SECTION TWO: REGULATION

Abbott, C. 2001. *Greater Portland: Urban Life and Landscape in the Pacific Northwest*. Philadelphia: University of Pennsylvania Press.

AFPC (Acting Food Policy Council). 2009. *Strategic Planning Framework*. Seattle: Acting Food Policy Council, Seattle King County. http://kitsapfoodchain.org/wp-content/uploads/2010/11/AFPC_Strategic-Framework_051209_FINAL.pdf.

Arroyo-Rodriguez, A., and C. Germain. 2012. "Ohio Guidance: Zoning Model for Urban Agriculture and Composting." *BioCycle* 53 (7): 24–25.

Balmer, K., J. Gill, H. Kaplinger, J. Miller, M. Peterson, A. Rhoads, P. Rosenbloom, and T. Wall. 2005. *The Diggable City: Making Urban Agriculture a Planning Priority*. Portland, OR: Portland State University.

Brinkley, C. 2013. "Avenues into Food Planning: A Review of Scholarly Food System Research." *International Planning Studies* 18 (2): 243–66.

Brinkley, C., E. Birch, and A. Keating. 2013. "Feeding Cities: Charting a Research and Practice Agenda toward Food Security." *Journal of Agriculture, Food Systems, and Community Development* 3 (4): 81–87. http://dx.doi.org/10.5304/jafscd.2013.034.008.

Brinkley, C., and D. Vitiello. 2013. "From Farm to Nuisance: Animal Agriculture and the Rise of Planning Regulation." *Journal of Planning History* 13 (2): 113–35.

Buzby, J. C., H. F. Wells, and J. Hyman. 2014. *The Estimated Amount, Value, and Calories of Postharvest Food Losses at the Retail and Consumer Levels in the United States* (EIB-121). US Department of Agriculture, Economic Research Service.

CIP (Cascadian Independence Project). 2013. "Cascadian Independence Project." www.cascadianow.org.

City of Austin, TX. 2008. *Austin, Texas Zero Waste Strategic Plan*. December 4. Prepared by Gary Liss and Associates, Loomis, CA.

———. 2009. City Council Resolution no. 20090115-050. January 15.

———. 2011. *Austin Resource Recovery Master Plan*. December 15.

City of Boston, MA. 2013. Zoning Code, Article 89 (Urban Agriculture), Section 2.7 (Definitions – Composting).

City of Chicago, IL. 2014. Municipal Code, Title 17 (Chicago Zoning Ordinance), Section 17-9-0103.1.

City of Cincinnati, OH. 2010. Zoning Code, Section 1419-41 (Community Garden).

City of Cleveland, OH. 2007. Codified Ordinances, Chapter 336 (Urban Garden District), Section 336.05 (Supplemental Regulations).

City of Dayton, OH. 2010. Zoning Code, Section 150.420.1.5 (Supplemental District Regulations – Composting).

City of Madison, WI. (2013). Zoning Code, Section 28.032 (Residential District Uses), Section 28.061 (Mixed-Use and Commercial District Uses), Section 28.082 (Employment District Uses), Section 28.091 (Special District Uses).

City of Portland, OR. 2009. City Charter, Chapter 17.102, "Solid Waste and Recycling Collection." May 15.

———. 2009. *Climate Action Plan*. Portland: Bureau of Planning and Sustainability.

———. 2012. *Urban Food Zoning Code Update: Enhancing Portlanders' Connection to Their Food and Community*. Portland: Bureau of Planning and Sustainability.

City of San Francisco, CA. 2002. Resolution for 75% Waste Diversion Goal (Resolution No. 679-02). September 30.

———. 2009. Mandatory Recycling and Composting (Ordinance No. 100-09). Amendment of the Whole in Board. June 9.

City of Seattle. 2008. *Local Food Action Initiative*.

———. 2012. *Food Action Plan*. Seattle: Office of Sustainability and the Environment.

City of Vancouver. 2011. *Greenest City Action 2020 Plan*.

———. 2013. *Vancouver Food Strategy*.

Cohen, S. 2013. Interview with the authors. Portland, OR. May 20.

Cooperband, L. 2002. *The Art and Science of Composting: A Resource for Farmers and Compost Producers*. University of Wisconsin–Madison Center for Integrated Agricultural Systems.

Covert, M., and A. Morales. 2014. "Successful Social Movement Organizing and the Formalization of Food Production." In *Informal City: Settings, Strategies, Responses*, edited by A. Loukaitou-Sideris and V. Mukhija. Cambridge, MA: MIT Press.

Cronon, W. 1991. *Nature's Metropolis: Chicago and the Great West*. New York: W. W. Norton.

Drake, L. 2014. "Governmentality in Urban Food Production? Following 'Community' from Intentions to Outcomes." *Urban Geography* 35 (2): 177–96.

Evans-Cowley, J., and A. Arroyo-Rodriguez. 2013. "Integrating Food Waste Diversion into Food Systems Planning: A Case Study of the Mississippi Gulf Coast." *Journal of Agriculture, Food Systems, and Community Development* 3 (3): 167–85.

Gottlieb, R., and A. Joshi. 2010. *Food Justice*. Cambridge, MA: MIT Press.

Gunders, D. 2012. *Wasted: How America Is Losing Up to 40 Percent of Its Food from Farm to Fork to Landfill*. IP: 12-06-B. Natural Resources Defense Council. August.

Hackworth, J. 2007. *The Neoliberal City: Governance, Ideology, and Development in American Urbanism*. Ithaca, NY: Cornell University Press.

Harrison, E. Z., and T. L. Richard. 1992. "Municipal Solid Waste Composting: Policy and Regulation." *Biomass and Bioenergy* 3 (3–4): 127–43.

Helphand, K. I. 2006. *Defiant Gardens: Making Gardens in Wartime*. San Antonio:

Trinity University Press.

Hodgson, K., M. C. Campbell, and M. Bailkey. 2011. *Urban Agriculture: Growing Healthy, Sustainable Places*. American Planning Association, Planning Advisory Service Report.

Horst, M. 2008. *Growing Green: An Inventory of Public Lands Suitable for Gardening in Seattle, Washington*. Seattle: University of Washington Press.

———. 2012. *A Scan of Existing City and County Plans and Policies*. Presentation to Puget Sound Regional Food Policy Council. http://www.psrc.org/assets/6830/rfpc_101411_policy_scan.pdf?processed=true.

Hou, J., J. Johnson, and L. Lawson. 2009. *Greening Cities, Growing Communities: Learning from Seattle's Urban Community Gardens*. Seattle: University of Washington Press.

Hynes, H. P. 1996. *A Patch of Eden: America's Inner City Gardeners*. White River Junction, VT: Chelsea Green Publishing.

Kaethler, T. M. 2006. *Growing Space: The Potential of Urban Agriculture in the City of Vancouver*. Vancouver: University of British Columbia Press.

Kneebone, E., and A. Berube. 2013. *Confronting Suburban Poverty in America*. Washington, DC: Brookings Institution Press.

Lake, A., and T. Townshend. 2006. "Obesogenic Environments: Exploring the Built and Food Environments." *Perspectives in Public Health* 126 (6): 262–67.

Lawson, L. J. 2005. *City Bountiful: A Century of Community Gardening in America*. Berkeley: University of California Press.

Levensto, M. 2012. Interview with the authors. Vancouver, BC. October 5.

Mansfield, B. 2013. Interview with the authors. Portland, OR. May 20.

Martin, R., and T. Marsden. 1999. "Food for Urban Spaces: The Development of Urban Food Production in England and Wales." *International Planning Studies* 4 (3): 389–412.

Meggs, G. 2010. "Strathcona Community Garden's 25th Anniversary Gives a Glimpse of the Future." http://www.geoffmeggs.ca/2010/07/11/strathcona-community-gardens-25th-anniversary-gives-a-glimpse-of-the-future/.

Morales, A., and G. Kettles. 2009. "Zoning for Markets and Street Merchants." *Zoning Practice* 25 (1): 1–8.

Mukherji, N., and A. Morales. 2010. "Zoning for Urban Agriculture." *Zoning Practice* 26 (3): 1–8.

Multnomah County. 2010. *Multnomah Food Action Plan*. Portland, OR: Multnomah County Office of Sustainability.

Murphy, K. 2014. "Farm-to-Table Living Takes Root." *New York Times*, March 11.

http://www.nytimes.com/2014/03/12/dining/farm-to-table-living-takes-root.html?_r=0.

Neuner, K., S. Kelly, and S. Raja. 2011. *Planning to Eat? Innovative Local Government Plans and Policies to Build Healthy Food Systems in the United States.* Buffalo: Food Systems Planning and Healthy Communities Lab, State University of New York.

Ohio Environmental Protection Agency. 2012. *Urban Agriculture, Composting and Zoning: A Zoning Code Model for Promoting Composting and Organic Waste Diversion through Sustainable Urban Agriculture.* GD# 1011. Columbus, OH: Division of Materials and Waste Management.

Organic Waste Systems. 2012. *Anaerobic Digester Feasibility Study: Executive Summary.* Report to the city of Madison, WI. https://www.cityofmadison.com/streets/compost/documents/FeasabilityReportexecutivesummary.pdf.

Permien, B. 2001. "The Growing Culture." *Pacific Rim Magazine.* http://langaraprm.com/2001/04/the-growing-culture/.

Plat, B., R. Ross, and M. Poland. 2012. "Promoting the Practice: Supportive Rules for Small-Scale Composting." *BioCycle* 53 (6): 21–24.

Pohl, L. 2013. Interview with the authors. Portland, OR. May 19.

Proceedings of the Third National Conference on City Planning. 1911. Cambridge: Cambridge University Press.

Purman, J. 2008. "Compost Rules: Regulating Source Separated Organics Composting Sites." *BioCycle* 49 (5): 42–45.

Robbins, W. G., and B. Katrine. 2011. *Nature's Northwest: The North Pacific Slope in the Twentieth Century.* Tucson: University of Arizona Press.

Runyon, L. 2013. "Forget Golf Courses: Subdivisions Draw Residents with Farms." National Public Radio, December 17. http://www.npr.org/sections/thesalt/2013/12/17/251713829/forget-golf-courses-subdivisions-draw-residents-with-farms.

Sanders, J. C. 2010. *Seattle and the Roots of Urban Sustainability.* Pittsburgh: University of Pittsburgh Press.

Seattle City Council. 2008. *Local Food Action Initiative.* Resolution 31019.

State of Illinois. 2013. Environmental Protection Act, Section 3.330 Pollution Control Facility (HB2335).

State of Massachusetts. n.d. Massachusetts Department of Agricultural Resources 330 CMR 25.00 (Agricultural Compost Program).

State of Ohio. 2012. Ohio Administrative Code, Chapter 3745-560 (Composting Regulations). Ohio Environmental Protection Agency, Division of Materials and Waste Management.

State of Wisconsin. 2012. Wisconsin Administrative Code, Chapter NR 502.12 (Yard, farm, food residuals and source-separated compostable material composting facilities).

Stocker, L., and K. Barnett. 1998. "The Significance and Praxis of Community-Based Sustainability Projects: Community Gardens in Western Australia." *Local Environment* 3 (2): 179–89.

UrbanAgLaw.org. 2013. "Homeowners Associations and Urban Ag." http://www.urbanaglaw.org/homeowners-associations/.

US Census Bureau. 2013. "2010 Census Urban and Rural Classification and Urban Area Criteria." July 22. http://www.census.gov/geo/reference/ua/urban-rural-2010.html.

US Environmental Protection Agency. 2013a. *History of the Resource Conservation and Recovery Act (RCRA)*. https://www.epa.gov/rcra/history-resource-conservation-and-recovery-act-rcra.

——— 2013b. *Municipal Solid Waste Generation, Recycling, and Disposal in the United States: Facts and Figures for 2011*. OIG publication no. EPA-530-F-13-001. Washington, DC: Solid Waste and Emergency Response. http://www.epa.gov/osw/nonhaz/municipal/msw99.htm.

——— 2013c. *Wastes – Laws & Regulations – History of RCRA*. https://www.epa.gov/rcra/resource-conservation-and-recovery-act-rcra-regulations#nonhaz.

———. 2014a. *Draft U.S. Greenhouse Gas Inventory Report: 1990–2014*. Washington, DC: US Environmental Protection Agency. http://www.epa.gov/climatechange/ghgemissions/usinventoryreport.html.

——— 2014b. *Municipal Solid Waste*. https://www3.epa.gov/epawaste/nonhaz/municipal/index.htm.

——— 2014c. *Municipal Solid Waste Generation, Recycling, and Disposal in the United States: Facts and Figures for 2012*. OIG publication no. EPA-530-F-14-001. Washington, DC: Solid Waste and Emergency Response. www.epa.gov/osw/nonhaz/municipal/msw99.htm.

——— 2014d. *Sustainable Management of Food*. http://www.epa.gov/composting/basic.htm.

Vitiello, D., and C. Brinkley. 2013. "The Hidden History of Food System Planning." *Journal of Planning History* 13 (2): 91–112. doi:10.1177/1538513213507541.

Woodsworth, B., and M. Levenston. 1979. "City Farmer's Vision of Urban Agriculture." International Science Education Symposium, University of British Columbia. http://cityfarmer.org/CFgoals1979.html#goals.

SECTION THREE: PRODUCTION

Ackerman, K. 2011. *The Potential for Urban Agriculture in New York City: Growing Capacity, Food Security, and Green Infrastructure*. New York: Urban Design Lab, Earth Institute, Columbia University.

Airriess, C. A., and D. L. Clawson. 1994. "Vietnamese Market Gardens in New Orleans." *Geographical Review* 84 (1): 16. doi:10.2307/215778.

Alaimo, K., E. Packnett, R. A. Miles, and D. J. Kruger. 2008. "Fruit and Vegetable Intake among Urban Community Gardeners." *Journal of Nutrition Education and Behavior* 40 (2): 94–101.

Alberts, H. R. 2014. "Development Battles." *Curbed New York*. http://ny.curbed.com/archives/2014/04/07/development_battles.php.

Altieri, M. A. 2014. "Agroecology: Principles and Strategies for Designing Sustainable Farming Systems." *Agroecology in Action*. Accessed July 14. http://nature.berkeley.edu/~miguel-alt/principles_and_strategies.html.

Aminyar, Y., and N. Fitzgerald. 2010. *New Brunswick Community Farmers Market Customer Survey Report*. New Brunswick, NJ: Rutgers University.

Ard, J. D., S. Fitzpatrick, R. A. Desmond, B. S. Sutton, M. Pisu, D. B. Allison, F. Franklin, and M. L. Baskin. 2007. "The Impact of Cost on the Availability of Fruits and Vegetables in the Homes of Schoolchildren in Birmingham, Alabama." *American Journal of Public Health* 97 (2): 367–72. http://doi.org/10.2105/AJPH.2005.080655.

Armstrong, D. 2000. "A Survey of Community Gardens in Upstate New York: Implications for Health Promotion and Community Development." *Health and Place* 6 (4): 319–27. doi:10.1016/S1353-8292(00)00013-7.

Arnstein, S. R. 1969. "A Ladder of Citizen Participation." *Journal of the American Institute of Planning* 35 (4): 216–24.

Astrup, A., J. Dyerbery, M. Selleck, and S. Stender. 2007. "Nutrition Transition and Its Relationship to the Development of Obesity and Related Chronic Diseases." *Obesity Review* 9 (Suppl. 1): 48–52.

Atiyeh, R. M., S. Subler, C. A. Edwards, G. Bachman, J. D. Metzger, and W. Shuster. 2000. "Effects of Vermicomposts and Composts on Plant Growth in Horticultural Container Media and Soil." *Pedobiologia* 44 (5): 579–90.

Baker, L. 2004. "Tending Cultural Landscapes and Food Citizenship in Toronto's Community Gardens." *Geographical Review* 94 (3): 305–25.

Balmer, K., J. Gill, H. Kaplinger, J. Miller, M. Peterson, A. Rhoads, P. Rosenbloom, and T. Wall. 2005. "The Diggable City: Making Urban Agriculture a Planning Priority." Master of Urban and Regional Planning Workshop Projects. http://

pdxscholar.library.pdx.edu/usp_murp/52.

Blair, D., C. C. Giesecke, and S. Sherman. 1991. "A Dietary, Social and Economic Evaluation of the Philadelphia Urban Gardening Project." *Journal of Nutrition Education* 23 (4): 161–67.

Blokland, T., 2012. "Blaming Neither the Undeserving Poor Nor the Revanchist Middle Classes: A Relational Approach to Marginalization." *Urban Geography* 33 (4): 488–507.

Brown, K. H., and A. L. Jameton. 2000. "Public Health Implications of Urban Agriculture." *Journal of Public Health Policy* 21 (1): 20–39.

Capece, A., M. Cassidy, and M. Sarsycki. 2012. *High Tunnels in New Brunswick*. Camden, NJ: Community Development Studio, Edward J. Bloustein School of Planning and Public Policy, Rutgers University.

Carey, E. E., L. Jett, W. J. Lamont, T. T. Nennich, M. D. Orzolek, and K. A. Williams. 2009. "Horticultural Crop Production in High Tunnels in the United States: A Snapshot." *HortTechnology* 19 (1): 37–43.

Companion, M. 2008. *An Overview of the State of Native American Health Challenges and Opportunities*. Washington, DC: International Relief and Development.

———. 2012. "Urban and Peri-Urban Cultivation in Northern Mozambique: Impacts on Food Security among Female Street Food Vendors." *Journal of Applied Social Science* 6:149–64.

———. 2013a. "Lessons from the 'Bucket Brigade': Using Social Ecology and Empowerment Models to Address Nutritional Education and Cultural Invigoration among Urban Native American Adults." *Indigenous Policy Journal* 24 (1): 1–16.

———. 2013b. "Obesogenic Cultural Drift and Nutritional Transition: Identifying Barriers to Healthier Food Consumption in Urban Native American Populations." *Journal of Applied Social Science* 7 (1): 80–94.

———. 2014. "Buckets of Fun!: Empowering Low-Income Urban Native American Youth to Make Nutritional Changes through Container Gardening." *Indigenous Policy Journal* 24 (3): 1–16.

Compher, C. 2006. "The Nutrition Transition in American Indians." *Journal of Transcultural Nursing* 13: 217–23.

Conway, G. R. 1985. "Agroecosystem Analysis." *Agricultural Administration* 20 (1): 31–55.

Corbin, J., and A. Strauss. 2008. *Basics of Qualitative Research: Techniques and Procedures for Developing Grounded Theory*. Los Angeles: Sage Publications.

Corrigan, M. P. 2011. "Growing What You Eat: Developing Community Gardens in

Baltimore, Maryland." *Applied Geography* 31 (4): 1232–41.

Covert, M. 2012. "Growing the Desert: Urban Agriculture Land Use Policy in the American West." Master's thesis, University of Wisconsin–Madison. http://digital.library.wisc.edu/1793/61971.

Covert, M., and A. Morales. 2014. "Successful Social Movement Organizing and the Formalization of Food Production." In *Informal City: Settings, Strategies, Responses*, edited by A. Loukaitou-Sideris and V. Mukhija. Cambridge, MA: MIT Press.

Damman, S., W. B. Eide, and H. V. Kuhnlein. 2008. "Indigenous Peoples' Nutrition Transition in a Right to Food Perspective." *Food Policy* 33:135–55.

Delormier, T., K. L. Frohlich, and L. Potvin. 2009. "Food and Eating as Social Practice: Understanding Eating Patterns as Social Phenomena and Implications for Public Health. *Sociology of Health and Illness* 31 (2): 215–28.

Drake, L. 2014. "Governmentality in Urban Food Production? Following 'Community' from Intentions to Outcome." *Urban Geography* 35 (2): 177–96.

Drake, L., and L. J. Lawson. 2015. "Results of a U.S. and Canada Community Garden Survey: Shared Challenges in Garden Management amid Diverse Geographical and Organizational Contexts." *Agriculture and Human Values* 32 (2): 241–54.

Fitzgerald, N. 2010. "What the Residents Have Told Us: Food Insecurity and Barriers to Fruit and Vegetable Intake in New Brunswick." Paper presented at the Learn, Grow, Share Community Forum, New Brunswick, NJ.

Fitzgerald, N., N. Czarnecki, and W. Hallman. 2010. "Food Intake Patterns and Perceived Barriers to Fruit and Vegetable Consumption among Minorities." Paper presented at the American Public Health Association Annual Meeting and Exposition, Denver, CO.

Fitzgerald, N., and N. Shah. 2010. *New Brunswick Community Farmers Market Vendor Survey Report*. New Brunswick, NJ: Rutgers University.

Flachs, A. 2010. "Food for Thought: The Social Impact of Community Gardens in the Greater Cleveland Area." *Electronic Green Journal* 1 (30): article 3. http://escholarship.org/uc/item/6bh7j4z4#page-1.

Foley, W. 2005. "Tradition and Change in Urban Indigenous Food Practice." *Postcolonial Studies* 8 (1): 25–44.

Francis, C., G. Lieblein, H. Steinsholt, T. A. Breland, J. Helenius, N. Sriskandarajah, and L. Salomonsson. 2005. "Food Systems and Environment: Building Positive Rural-Urban Linkages." *Human Ecology Review* 12 (1): 60–71.

Freeman, J. 1972. "Tyranny of Structurelessness." *Berkeley Journal of Sociology* 17:151–64.

Frohlich, K. L., E. Corin, and L. Potvin. 2001. "A Theoretical Proposal for the Relationship between Context and Disease." *Sociology of Health and Illness* 23 (6): 776–97.

Gittelsohn, J., J. A. Anliker, S. Sharma, A. E. Vastine, B. Caballero, and B. Ethelbah. 2006. "Psychosocial Determinants of Food Purchasing and Preparation in American Indian Households." *Journal of Nutritional Education* 38:163–68.

Glover, T. D., K. J. Shinew, and D. C. Parry. 2005. "Association, Sociability, and Civic Culture: The Democratic Effect of Community Gardening." *Leisure Sciences* 27 (1): 75–92. doi:10.1080/01490400590886060.

Goldstein, M., J. Bellis, S. Morse, A. Myers, and E. Ura. 2011. *Urban Agriculture—A Sixteen City Survey of Urban Agriculture Practices across the Country*. Turner Environmental Law Clinic, Emory Law School. http://www.jhsph.edu/research/centers-and-institutes/johns-hopkins-center-for-a-livable-future/_pdf/projects/FPN/Urban_Community_Planning/URBAN_AGRICULTURE_A_SIXTEENCITY_SURVEY_OF_URBAN_AGRICULTURE_PRACTICES_ACROSS_THE_COUNTRY.pdf.

Gottlieb, R., and A. Joshi. 2010. *Food Justice*. Cambridge, MA: MIT Press.

Guarnaccia, P., T. Vivar, A. C. Bellows, and G. Alcarez. 2012. "'We Eat Meat Every Day': Ecology and Economy of Dietary Change among Oaxacan Migrants from Mexico to New Jersey." *Ethnic and Racial Studies* 35 (1): 104–9.

Guthman, J. 2008. "Bringing Good Food to Others: Investigating the Subjects of Alternative Food Practice." *Cultural Geographies* 15 (4): 431–47. doi:10.1177/1474474008094315.

Hagey, A., S. Rice, and R. Flournoy. 2012. *Growing Urban Agriculture: Equitable Strategies and Policies for Improving Access to Healthy Food and Revitalizing Communities*. PolicyLink. https://www.policylink.org/sites/default/files/URBAN_AG_FULLREPORT.PDF.

Hale, J., C. Knapp, L. Bardwell, M. Buchenau, J. Marshall, F. Sancar, and J. S. Litt. 2011. "Connecting Food Environments and Health through the Relational Nature of Aesthetics: Gaining Insight through the Community Gardening Experience." *Social Science and Medicine*. doi:10.1016/j.socscimed.2011.03.044.

Halpern, P. 2007. *Obesity and American Indians/Alaska Natives*. Washington, DC: US Department of Health and Human Services, Office of the Assistant Secretary for Planning and Evaluation.

Hamm, M. W., and A. C. Bellows. 2003. "Community Food Security and Nutrition Educators." *Journal of Nutrition Education and Behavior* 35 (1): 37–43.

Hanna, A. K., and P. Oh. 2000. "Rethinking Urban Poverty: A Look at Community

Gardens." *Bulletin of Science, Technology and Society* 20:207–16.

Harris, E. 2009. "Neoliberal Subjectivities or a Politics of the Possible? Reading for Difference in Alternative Food Networks." *Area* 41 (1): 55–63.

Hawkes, C. 2008. "Dietary Implications of Supermarket Development: A Global Perspective." *Development Policy Review* 26 (6): 657–92.

Hendrickson, M. K., and M. Porth. 2012. *Urban Agriculture—Best Practices and Possibilities*. University of Missouri Extension. http://extension.missouri.edu/foodsystems/documents/urbanagreport_072012.pdf.

Hsieh, P-L. 2004. "Factors Influencing Students' Decision to Choose Healthy or Unhealthy Snacks at the University of Newcastle, Australia." *Journal of Nursing Research* 12 (2): 83–91.

Hu, A., A. Acosta, A. McDaniel, and J. Gittelsohn. 2013. "Community Perspectives on Barriers and Strategies for Promoting Locally Grown Produce from an Urban Agriculture Farm." *Health Promotion Practice* 14 (1): 69–74.

Huntsberry, W. 2014. "New Garden in the Bronx 'Creates a Community.'" *Wall Street Journal*, June 18. http://online.wsj.com/articles/new-garden-in-the-bronx-creates-a-community-1403140376.

James, D. C. 2004. "Factors Influencing Food Choices, Dietary Intake, and Nutrition-Related Attitudes among African Americans: Application of a Culturally Sensitive Model." *Ethnicity and Health* 9 (4): 349–67.

Johnson & Johnson. n.d. "Johnson & Johnson History." http://www.jnj.com/about-jnj/company-history.

———. n.d. "Our Community Work in New Jersey." http://www.jnj.com/sites/default/files/pdf/local-community-responsibility.pdf.

Kaufman, J., and M. Bailkey. 2000. *Farming inside Cities: Entrepreneurial Urban Agriculture in the United States*. Cambridge, MA: Lincoln Institute of Land Policy.

Kellogg Commission on the Future of the State and Land Grant Universities. 1999. *Returning to Our Roots: The Engaged Institution*. Third report. Washington, DC: National Association of State Universities and Land Grant Colleges.

Kennedy, S. H. 2009. *The Garden*. Documentary. http://www.thegardenmovie.com/.

Kingsley, J. Y., M. Townsend, and C. Henderson-Wilson. 2009. "Cultivating Health and Wellbeing: Members' Perceptions of the Health Benefits of a Port Melbourne Community Garden." *Leisure Studies* 28 (2): 207–19.

Kobayashi, M., L. Tyson, and J. Abi-Nader. 2010. *The Activities and Impacts of Community Food Projects 2005–2009*. Report from the Community Food Project Competitive Grants Program, 1–28. Community Food Security Coalition, US Department of Agriculture, National Research Institute.

Krasny, M., and R. Doyle. 2002. "Participatory Approaches to Program Development and Engaging Youth in Research: The Case of an Intergenerational Urban Community Gardening Program." *Journal of Extension* 40 (5): 3–10.

Krasny, M., and K. Tidball. 2009. "Community Gardens as Contexts for Science, Stewardship, and Civic Action Learning." *Cities and the Environment* 2 (1): 1–18.

Kuhnlein, H. V., and O. Receveur. 2007. "Local Cultural Animal Food Contributes High Levels of Nutrients for Arctic Canadian Indigenous Adults and Children." *Journal of Nutrition* 137:1110–14.

Kurtz, H. E. 2001. "Differentiating Multiple Meanings of Garden and Community." *Urban Geography* 22:656–70.

Lake, A., and T. Townshend. 2006. "Obesogenic Environments: Exploring the Built and Food Environments." *Perspectives in Public Health* 126 (6): 262–67.

Larsen, K., and J. Gilliland. 2009. "A Farmers' Market in a Food Desert: Evaluating Impacts on the Price and Availability of Healthy Food." *Health and Place* 15 (4): 1158–62.

Larson, N., and M. Story. 2009. "A Review of Environmental Influences on Food Choices." *Annals of Behavioral Medicine* 38 (Supp. 1): S56–S73.

Lawson, L. J. 2005. *City Bountiful: A Century of Community Gardening in America.* Berkeley: University of California Press.

Litt, J. S., M-J. Soonbader, M. S. Turbin, J. W. Hale, M. Buchenau, and J. A. Marshall. 2011. "The Influence of Social Involvement, Neighborhood Aesthetics, and Community Garden Participation on Fruit and Vegetable Consumption." *American Journal of Public Health* 101 (8): 1466–73.

Lloyd, K., P. Ohri-Vachaspati, S. Brownlee, M. Yedidia, D. Gaboda, and J. Chou. 2010. *New Jersey Childhood Obesity Survey.* Chartbook. New Brunswick, NJ: Rutgers Center for State Health Policy.

Mayfield, L., and E. Lucas. 2000. "Mutual Awareness, Mutual Respect: The Community and the University Interact." *Cityscape: A Journal of Policy Development and Research* 5 (1): 173–84.

McCormack, L. A., M. Story, N. I. Larson, and M. N. Laska. 2010. "Review of the Nutritional Implications of Farmers' Markets and Community Gardens: A Call for Evaluation and Research Efforts." *Journal of the American Dietetic Association* 110 (3): 399–408. doi:http://dx.doi.org/10.1016/j.jada.2009.11.023.

Milburn, L.-A. S., and B. A. Vail. 2010. "Sowing the Seeds of Success: Cultivating a Future for Community Gardens." *Landscape Journal* 29 (1): 71–89.

Miller, R. W. 1997. *Urban Forestry: Planning and Managing Urban Greenspaces.* Upper

Saddle River, NJ: Prentice Hall.
Mogk, J. E., S. Wiatkowski, and M. J. Weindorf. 2010. "Promoting Urban Agriculture as an Alternative Land Use for Vacant Properties in the City of Detroit: Benefits, Problems and Proposals for a Regulatory Framework for Successful Land Use Integration." *Wayne Law Review* 56:1521.
Mukherji, N., and A. Morales. 2010. "Zoning for Urban Agriculture." *Zoning Practice*, March. www.planning.org/zoningpractice/2010/pdf/mar.pdf.
New York City Council. 2010. *FoodWorks: A Vision to Improve NYC's Food System*. http://council.nyc.gov/downloads/pdf/foodworks_fullreport_11_22_10.pdf.
Obeng-Odoom, F. 2013. "Underwriting Food Security in the Urban Way: Lessons from African Countries." *Agroecology and Sustainable Food Systems* 37:614–28.
Ober Allen, J., K. Alaimo, D. Elam, and E. Perry. 2008. "Growing Vegetables and Values: Benefits of Neighborhood-Based Community Gardens for Youth Development and Nutrition." *Journal of Hunger and Environmental Nutrition* 3 (4): 418–39.
Ostrander, S. A. 2004. "Democracy, Civic Participation, and the University: A Comparative Study of Civic Engagement on Five Campuses." *Nonprofit and Voluntary Sector Quarterly* 33 (1): 74–93.
Oxenham, E., and A. D. King. 2010. "School Gardens as a Strategy for Increasing Fruit and Vegetable Consumption." *School Nutrition Association* 34 (1).
Patel, I. C. 1996. "Rutgers Urban Gardening: A Case Study in Urban Agriculture." *Journal of Agricultural and Food Information* 3 (3): 35–46.
Payne, K., and D. Fryman. 2001. *Cultivating Community: Principles and Practices for Community Gardening as a Community Building Tool*. College Park, GA: American Community Gardening Association.
Pearce, J. L. 1993. *Volunteers: The Organizational Behavior of Unpaid Workers*. London: Routledge.
Petersen, E. 2011. "Urban Gardens Lead the Way for Water Policy in Milwaukee." http://urbanmilwaukee.com/2011/04/15/urban-gardens-lead-the-way-for-water-policy-in-milwaukee/.
Pfeiffer, A., E. Silva, and J. Colquhoun. 2014. "Innovation in Urban-Agriculture Practices: Responding to Diverse Production Environments." *Renewable Agriculture and Food Systems* 30 (Special Issue 1): 79–91.
Popkin, B. M. 2004. "The Nutrition Transition: An Overview of World Patterns of Change." *Nutrition Reviews* 62 (7): S140–S143.
Pothukuchi, K., and J. Kaufman. 1999. "Placing the Food System on the Urban Agenda: The Role of Municipal Institutions in Food Systems Planning."

Agriculture and Human Values 16 (2): 213–24.

———. 2000. "The Food System: A Stranger to Planning." *APA Journal* 66 (2): 113–24.

Prospect Heights Community Farm. 2014. "Get Involved: New Members." Accessed June 23. http://www.phcfarm.com/welcome/garden-member-info/new-members/.

Raja, S., B. Born, and J. K. Russell. 2008. *A Planners Guide to Community and Regional Food Planning*. Planning Advisory Service Report Number 554. Chicago: American Planning Association.

Raschke, V., and B. Cheema. 2007. "Colonization, the New World Order, and the Eradication of Traditional Food Habits in East Africa: Historical Perspective on the Nutrition Transition." *Public Health Nutrition* 11 (7): 662–74.

Reardon, K. M. 1996. "Town/Gown Conflicts: Campus/Community Partnerships in the 90s, with Thomas P. Shields." *Planners Network Newsletter*, March, 10–11.

Reuther, S., and N. Dewar. 2005. "Competition for the Use of Public Open Space in Low-Income Urban Areas: The Economic Potential of Urban Gardening in Khayelitsha, Cape Town." *Development Southern Africa* 23 (1): 97–122.

Ruddick, S. 1996. "Constructing Difference in Public Spaces: Race, Class, and Gender as Interlocking Systems." *Urban Geography* 17 (2): 132–51.

Saldivar-Tanaka, L., and M. E. Krasny. 2004. "Culturing Community Development, Neighborhood Open Space, and Civic Agriculture: The Case of Latino Community Gardens in New York City." *Agriculture and Human Values* 21 (4): 399–412.

Saul, N., and A. Curtis. 2013. *The Stop: How the Fight for Good Food Transformed a Community and Inspired a Movement*. Brooklyn, NY: Melville House.

Schmelzkopf, K. 1995. "Urban Community Gardens as Contested Space." *Geographical Review* 85 (3): 364. doi:10.2307/215279.

Skoglund, A. G. 2006. "Do Not Forget about Your Volunteers: A Qualitative Analysis of Factors Influencing Volunteer Turnover." *Health and Social Work* 31 (3): 217–20.

Sorensen, J., and L. Lawson. 2012. "Evolution in Partnership: Lessons from the East St. Louis Action Research Project." *Action Research Journal* 10 (2): 150–69.

Spalding, B., N. Czarnecki, W. Hallman, and N. Fitzgerald. 2012. "Can Farmers Markets Improve Access and Consumption of Fruits and Vegetables in Vulnerable Populations?" Paper presented at the Academy of Nutrition and Dietetics Food and Nutrition Conference and Expo (FNCE), Philadelphia, PA. Abstract: *Journal of the Academy of Nutrition and Dietetics* 112 (Suppl. 3): A-72.

Stevenson, C., G. Doherty, J. Barnett, M. Muldoon, and K. Trew. 2007. "Adolescents' Views of Food and Eating: Identifying Barriers to Healthy Eating." *Journal of Adolescence* 30: 417–34.

Stoecker, R. 1999. "Are Academics Irrelevant? Rules for Scholars in Participatory Research." *American Behavioral Scientist* 42 (5): 840–54.

Teig, E., J. Amulya, L. Bardwell, M. Buchenau, J. A. Marshall, and J. S. Litt. 2009. "Collective Efficacy in Denver, Colorado: Strengthening Neighborhoods and Health through Community Gardens." *Health and Place* 15 (4): 1115–22.

Twiss, J., J. Dickinson, S. Duma, T. Kleinman, H. Paulsen, and L. Rilveria. 2003. "Community Gardens: Lessons Learned from California Healthy Cities and Communities." *American Journal of Public Health* 93 (9): 1435–38. doi:10.2105/AJPH.93.9.1435.

US Census Bureau. 2010. "Quick Facts: New Brunswick City, New Jersey." http://www.census.gov/quickfacts/table/PST045215/3451210/ql.

USDA (US Department of Agriculture). 2014. "USDA Announces $78 Million Available for Local Food Enterprises." http://www.usda.gov/wps/portal/usda/usdahome?contentid=2014/05/0084.xml.

Weiner, M. D., T. D. MacKinnon, and O. T. Puniello. 2011. *New Brunswick Tomorrow 2011 Needs Assessment*. New Brunswick, NJ: Bloustein Center for Survey Research, Rutgers University.

Wiig, K., and C. Smith. 2008. "The Art of Grocery Shopping on a Food Stamp Budget: Factors Influencing the Food Choices of Low-Income Women as They Try to Make Ends Meet." *Public Health Nutrition* 12 (10): 1726–34.

Winne, M. 2008. *Closing the Food Gap: Resetting the Table in the Land of Plenty*. Boston: Beacon Press.

Zenk, S. N., A. J. Schulz, T. Hollis-Neely, R. T. Campbell, N. Holmes, G. Watkins, R. Nwankwo, and A. Odoms-Young. 2005a. "Fruit and Vegetable Intake in African Americans: Income and Store Characteristics." *American Journal of Preventive Medicine* 29 (1): 1–9.

Zenk, S. N., A. J. Schulz, B. A. Israel, S. A. James, S. Bao, and M. L. Wilson. 2005b. "Neighborhood Racial Composition, Neighborhood Poverty, and the Spatial Accessibility of Supermarkets in Metropolitan Detroit." *American Journal of Public Health* 95 (4): 660–67.

SECTION FOUR: DISTRIBUTION

Abatekassa, G., and H. C. Peterson. 2011. "Market Access for Local Food through the Conventional Food Supply Chain." *International Food and Agribusiness*

Management Review 14:63–82.

Ackerman, K. 2011. *The Potential for Urban Agriculture in New York City: Growing Capacity, Food Security, and Green Infrastructure.* New York: Urban Design Lab, Earth Institute, Columbia University.

Ahmed, A., and D. Little. 2008. "Chicago, America's Most Segregated Big City." *Chicago Tribune.* December 26. http://articles.chicagotribune.com/2008-12-26/news/0812250194_1_racial-steering-douglas-massey-neighborhoods.

Alwitt, L. F., and T. D. Donley. 1997. "Retail Stores in Poor Urban Neighborhoods." *Journal of Consumer Affairs* 31 (1): 139–64.

Anderson, R. N., and B. L. Smith. 2003. *Deaths: Leading Causes for 2001.* Centers for Disease Control and Prevention, National Center for Health Statistics.

Andersonville Farmer's Market. 2012. Andersonville Chamber of Commerce. http://www.andersonville.org/events/andersonville-farmers-market.

Apparicio, P., M.-S. Cloutier, and R. Shearmur. 2007. "The Case of Montréal's Missing Food Deserts: Evaluation of Accessibility to Food Supermarkets." *International Journal of Health Geographics* 6:1–13. doi:10.1186/1476-072X-6-4.

Bader, M. D. M., M. Purciel, P. Yousefzadeh, and K. M. Neckerman. 2010. "Disparities in Neighborhood Food Environments: Implications of Measurement Strategies." *Economic Geography* 86:409–30.

Baker, D., K. Hamshaw, and J. Kolodinsky. 2009. "Who Shops at the Market? Using Consumer Surveys to Grow Farmers' Markets: Findings from a Regional Market in Northwestern Vermont." *Journal of Extension* 47:1–9.

Balkin, S., and B. Mier. 2001. "Maxwell Street: Chicago Illinois." In *Celebrating the Third Place: Inspiring Stories about the "Great Good Places" at the Heart of our Communities*, edited by R. Oldenburg, 193–208. New York: Marlowe.

Barham, J., D. Tropp, K. Enterline, J. Farbman, J. Fisk, and S. Kiraly. 2012. *Regional Food Hub Resource Guide.* US Department of Agriculture, Agricultural Marketing Service.

Bloom, J. D., and C. C. Hinrichs. 2011. "Informal and Formal Mechanisms of Coordination in Hybrid Food Value Chains." *Journal of Agriculture, Food Systems, and Community Development* 1:143–56.

Boys, K. A., and D. H. Hughes. 2013. "A Regional Economics-Based Research Agenda for Local Food Systems." *Journal of Agriculture, Food Systems, and Community Development* 3:145–50.

Chicago Department of Public Health. 2011. "Transforming the Health of Our City: Chicago Answers the Call." http://www.cityofchicago.org/dam/city/depts/cdph/CDPH/PublicHlthAgenda2011.pdf.

Chung, C., and S. L. Myers. 1999. "Do the Poor Pay More for Food? An Analysis of Grocery Store Availability and Food Price Disparities." *Journal of Consumer Affairs* 33:276–96.

City of Chicago Census Maps. 2000. US Census. Updated 2013. http://www.cityofchicago.org/city/en/depts/doit/supp_info/census_maps.html.

Clancy, K., and K. Ruhf. 2010. "Regional Value Chains in the Northeast: Findings from a Survey." Northeast Sustainable Agriculture Working Group. http://nesawg.org/resources/regional-value-chains-northeast-findings-survey.

Clarke, G., H. Eyre, and C. Guy. 2002. "Deriving Indicators of Access to Food Retail Provision in British Cities: Studies of Cardiff, Leeds and Bradford." *Urban Studies* 39:2041–60. doi:10.1080/00420980220000011353.

Cohen, N., and D. Derryck. 2011. "Corbin Hill Road Farm Share: A Hybrid Food Value Chain in Practice." *Journal of Agriculture, Food Systems, and Community Development* 1:85–100.

Conner, D., W. Knudson, M. Hamm, and H. C. Peterson. 2008. "The Food System as an Economic Driver: Strategies and Applications for Michigan." *Journal of Hunger and Environmental Nutrition* 3:371–83.

Cronon, W. 1991. *Nature's Metropolis: Chicago and the Great West*. New York: W. W. Norton.

Cuomo, A. 2013. "Governor Cuomo Announces $3.6 Million in CFA Funding for Food Distribution Hubs." Albany, NY: Governor's Press Office.

Davis, J., D. Merriman, L. Samayoa, B. Flanagan, R. Baiman, and J. Persky. 2009. *The Impact of an Urban Wal-Mart Store on Area Businesses: An Evaluation of One Chicago Neighborhood's Experience*, 1–67. Chicago: Center for Urban Research and Learning, Loyola University. http://ecommons.luc.edu/cgi/viewcontent.cgi?article=1002&context=curl_pubs.

Day Farnsworth, L., and A. Morales. 2011. "Satiating the Demand: Planning for Alternative Models of Regional Food Distribution." *Journal of Agriculture, Food Systems, and Community Development* 2:227–47.

Deutsch, G. 1904. "Hawkers and Peddlers." In *The Jewish Encyclopedia*, edited by I. Singer, 267–69. New York: Funk and Wagnalls.

Diamond, A., and J. Barham. 2011. "Money and Mission: Moving Food with Value and Values." *Journal of Agriculture, Food Systems, and Community Development* 1:101–17.

Eisenhauer, E. 2001. "In Poor Health: Supermarket Redlining and Urban Nutrition." *GeoJournal* 53:125–33.

Evans, G. W. 2004. "The Environment of Childhood Poverty." *American Psychologist*

59 (2): 77.

Feenstra, G., P. Allen, S. Hardesty, J. Ohmart, and J. Perez. 2011. "Using a Supply Chain Analysis to Assess the Sustainability of Farm-to-Institution Programs." *Journal of Agriculture, Food Systems, and Community Development* 1:69–84.

Fischer, M., M. Hamm, R. Pirog, J. Fisk, J. Farbman, and S. Kiraly. 2013. *Findings of the 2013 National Food Hub Survey*. East Lansing: Michigan State University Center for Regional Food Systems.

Freedman, D. A., and B. A. Bell. 2009. "Access to Healthful Foods among an Urban Food Insecure Population: Perceptions versus Reality. *Journal of Urban Health* 86:825–38. doi:10.1007/s11524-009-9408-x.

Fried, B. 2005. "For the Health of It: Farmers Markets Boost the Prospects of Low-Income Communities with Fresh, Wholesome Food." Project for Public Spaces. Paper in possession of author.

Gallagher, M. 2006. *Examining the Impact of Food Deserts on Public Health in Chicago*. Mari Gallagher Research and Consulting Group. http://marigallagher.com/projects/4/.

———. 2010a. *Examining the Impact of Food Deserts and Food Imbalance on Public Health in Birmingham, Alabama*. Mari Gallagher Research and Consulting Group, 1–68. http://marigallagher.com/site_media/dynamic/project_files/Birm_Report_Cond.pdf

———. 2010b. *Examining the Impact of Food Deserts and Food Imbalance on Public Health in Birmingham, Alabama: Technical Appendix*. Mari Gallagher Research and Consulting Group, 1–21. http://marigallagher.com/site_media/dynamic/project_files/Birm_Appendix.pdf

———. 2011a. *Food and Health in Hamilton County, OH: Technical Report*. Mari Gallagher Research and Consulting Group, 1–79. http://marigallagher.com/site_media/dynamic/project_files/HamiltonCtyFdHlth.pdf

———. 2011b. *The Chicago Food Desert Progress Report*. Mari Gallagher Research and Consulting Group, 1–4. http://marigallagher.com/site_media/dynamic/project_files/FoodDesert2011.pdf

Gerend, J. 2007. "Temps Welcome." *Planning* 73:24–27.

Glanz, K., J. F. Sallis, B. E. Saelens, and L. D. Frank. 2007. "Nutrition Environment Measures Survey in Stores (NEMS-S): Development and Evaluation." *American Journal of Preventive Medicine* 32:282–89. doi:10.1016/j.amepre.2006.12.019.

Green City Market. 2012. "About Green City Market." http://www.greencitymarket.org/about/.

Guy, C., G. Clarke, and H. Eyre. 2004. "Food Retail Change and the Growth of Food Deserts: A Case Study of Cardiff." *International Journal of Retail and Distribution Management* 32:72.

Hallett, L. F., IV, and D. McDermott. 2011. "Quantifying the Extent and Cost of Food Deserts in Lawrence, Kansas, USA." *Applied Geography* 31:1210–15. doi:10.1016/j.apgeog.2010.09.006.

Hardesty, S. D., G. Feenstra, D. Visher, T. Lerman, D. Thilmany-McFadden, A. Bauman, T. Gillpatrick, and G. N. Rainbolt. 2014. "Values-Based Supply Chains: Supporting Regional Food and Farms." *Economic Development Quarterly* 28 (1): 17–27.

Hardesty, S. D.. 2008. "The Growing Role of Local Food Markets." *American Journal of Agricultural Economics* 90:1289–95.

Hardesty, S. D., and P. Leff. 2010. "Determining Marketing Costs and Returns in Alternative Marketing Channels." *Renewable Agriculture and Food Systems* 25 (1): 24–34.

Herries, J. 2010. ESRI, Redlands, CA. http://megacity.esri.com/fooddeserts.

Hinrichs, C. C., and L. Charles. 2012. "Local Food Systems and Networks in the US and UK." In *Rural Transformations and Rural Policies in the US and UK*, edited by M. Shucksmith, D. L. Brown, S. Shortall, J. Vergunst, and M. E. Warner, 156–76. New York: Routledge.

Hodgson, K., M. C. Campbell, and M. Bailkey. 2011. *Urban Agriculture: Growing Healthy, Sustainable Places*. American Planning Association, Planning Advisory Service Report.

Hoshide, A. K. 2007. *Values-Based and Value-Added Value Chains in the Northeast, Upper Midwest, and Pacific Northwest*. Orono: University of Maine.

Institute of Medicine. 2001. *Health and Behavior: The Interplay of Biological, Behavioral, and Societal Influences*. Washington, DC: National Academies Press.

Jablonski, B. B. R. 2014. "Evaluating the Impact of Farmers' Markets Using a Rural Wealth Creation Approach." Chap. 14 in *Rural Wealth Creation*, edited by J. Pender, T. Johnson, B. Weber, and J. M. Fannin. New York: Routledge.

Jablonski, B. B. R., J. Perez-Burgos, and M. I. Gómez. 2011. "Food Value Chain Development in Central New York: CNY Bounty." *Journal of Agriculture, Food Systems, and Community Development* 1:129–41.

Jarosz, L. 2000. "Understanding Agri-Food Networks as Social Relations." *Agriculture and Human Values* 17:279–83.

Johnson, R., R. A. Aussenberg, and T. Cowan. 2013. *The Role of Local Food Systems in U.S. Farm Policy*. CRP Report for Congress 7-5700. Congressional Research Service.

Kershaw, T., T. Creighton, J. Marko, and T. Markham. 2010. "Food Access in

Saskatoon." Public Health Services. https://www.saskatoonhealthregion
.ca/locations_services/Services/Health-Observatory/Documents/Reports-
Publications/CommunityFoodAccessReportOct2010.pdf.

King, R. P., M. S. Hand, G. DiGiacomo, K. Clancy, M. I. Gómez, S. D. Hardesty, L. Lev, and E. W. McLaughlin. 2010. *Comparing the Structure, Size, and Performance of Local and Mainstream Food Supply Chains*. Economic Research Report ERR-99. US Department of Agriculture, Economic Research Service.

King, R. P., and L. Venturini. 2005. *Demand for Quality Drives Changes in Food Supply Chains*. Agriculture Information Bulletin 794. US Department of Agriculture, Economic Research Service.

King, S. C., A. J. Weber, H. L. Meiselman, and N. Lv. 2004. "The Effect of Meal Situation, Social Interaction, Physical Environment and Choice on Food Acceptability." *Food Quality and Preference* 15 (7): 645–53.

Knack, M. C. 2005. "Women and Men." Chap. 3 in *A Companion to the Anthropology of American Indians*, edited by T. Biolsi, 51. Hoboken, NJ: Wiley-Blackwell.

LaFollette Park Farmers Market. 2012. Telephone interview with A. Roubal. September 4.

Larmer, M. 2012. Telephone interview with A. Roubal. August 28.

Leete, L., N. Bania, and A. Sparks-Ibanga. 2012. "Congruence and Coverage: Alternative Approaches to Identifying Urban Food Deserts and Food Hinterlands." *Journal of Planning Education and Research* 32:204–18. doi:10.1177/0739456X11427145.

Lerman, T. 2012a. *A Review of Scholarly Literature on Values-Based Supply Chains*. Davis: University of California Sustainable Agriculture Research and Education Program and the Agricultural Sustainability Institute.

———. 2012b. *Values-Based Supply Chain Annotated Bibliography*. Davis: University of California Sustainable Agriculture Research and Education Program and the Agricultural Sustainability Institute.

LeRoux, M. N., T. M. Schmit, M. Roth, and D. H. Streeter. 2010. "Evaluating Marketing Channel Options for Small-Scale Fruit and Vegetable Producers." *Renewable Agriculture and Food Systems* 25 (1): 16–23.

Low, S. A., and S. Vogel. 2011. *Direct and Intermediated Marketing of Local Foods in the United States*. Economic Research Report ERR-128. US Department of Agriculture, Economic Research Service.

Martinez, S., M. Hand, M. Da Pra, S. Pollack, K. Ralston, T. Smith, S. Vogel, S. Clark, L. Lohr, S. Low, and C. Newman. 2010. *Local Food Systems: Concepts, Impacts, and Issues*. Economic Research Report ERR-97. US Department of Agriculture,

Economic Research Service.

Matteson, G., C. Gerencer, and E. Pirro. 2013. *Pathways to Food Hub Success: Financial Benchmark Metrics and Measurements for Regional Food Hubs*. National Good Food Network (NGFN) Webinar. http://www.ngfn.org/resources/ngfn-cluster-calls/financial-benchmarks-for-food-hubs/Food%20Hubs%20Benchmark%20slides.pdf.

Mission of the 61st Street Farmers Market. 2012. http://experimentalstation.org/mission.

Morales, A. 2000. "Peddling Policy: Street Vending in Historical and Contemporary Context." *International Journal of Sociology and Social Policy* 20 (3/4): 76–98.

———. 2011. "Marketplaces: Prospects for Social, Economic, and Political Development." *Journal of Planning Literature* 26 (1): 3–17.

Morales, A., S. Balkin, and J. Persky. 1995. "The Value of Benefits of a Public Street Market: The Case of Maxwell Street." *Economic Development Quarterly* 9:304–20.

Morland, K., S. Wing, and A. Diez Roux. 2002. "The Contextual Effect of the Local Food Environment on Residents' Diets: The Atherosclerosis Risk in Communities Study." *American Journal of Public Health* 90:1761–68.

Moths, J. 2010. *Toward a Localized Analysis of Quality Food Access on Chicago's South Side*. http://jessimoths.com/documents/moths_ba_thesis.pdf.

NGFN (National Good Food Network). 2013. Food Hub Center. http://www.ngfn.org/resources/food-hubs.

Painter, K. 2007. "An Analysis of Food-Chain Demand for Differentiated Farm Commodities: Implications for the Farm Sector." Washington, DC: USDA Rural Business and Cooperatives Program.

Pirenne, H. 1925. *Medieval Cities: Their Origins and the Revival of Trade*. Princeton, NJ: Princeton University Press.

PolicyLink. 2013. *Equitable Development Toolkit: Urban Agriculture and Community Gardens*. https://www.policylink.org/sites/default/files/urban-agriculture_0.pdf.

Pollan, M. 2008. *In Defense of Food: An Eater's Manifesto*. New York: Penguin Books.

Richards, T. J., and G. Pofahl. 2010. "Pricing Power by Supermarket Retailers: A Ghost in the Machine?" *Choices: The Magazine of Food, Farm and Resource Issues* 25:1–12.

Richardson, A., J. Boone-Heinonen, B. M. Popkin, and P. Gordon-Larsen. 2012. "Are Neighborhood Food Resources Distributed Inequitably by Income and Race? Epidemiologic Findings across the Urban Spectrum." *British Medical Journal* 2:1–77.

Rose, J. M. 1970. "Direct Charge Cooperatives: Legal Aspects of a New Strategy in the War on Poverty." *George Washington Law Review* 38 (5): 958–73.

Santilli, A., A. Carroll-Scott, F. Wong, and J. Ickovics. 2011. "Urban Youths Go 3000 Miles: Engaging and Supporting Young Residents to Conduct Neighborhood Asset Mapping." *American Journal of Public Health* 101:2207–10. doi:10.2105/AJPH.2011.300351.

Schmidt, M. C., J. M. Kolodinsky, T. P. DeSisto, and F. C. Conte. 2011. "Increasing Farm Income and Local Food Access: A Case Study of a Collaborative Aggregation, Marketing, and Distribution Strategy That Links Farmers to Markets." *Journal of Agriculture, Food Systems, and Community Development* 1:157–75.

Schneider, M. L., and C. A. Francis. 2005. "Marketing Locally Produced Foods: Consumer and Farmer Opinions in Washington County, Nebraska." *Renewable Agriculture and Food Systems* 20:252–60.

Sexton, R. J. 2010. "Grocery Retailers' Dominant Role in Evolving World Food Markets." *Choices: The Magazine of Food, Farm and Resource Issues* 25:1–13.

Shaw, H. J. 2006. "Food Deserts: Towards the Development of a Classification." *Annals of the Association of American Geographers* 88B:231–47.

Smoyer-Tomic, K. E., J. C. Spence, and C. Amrhein. 2006. "Food Deserts in the Prairies? Supermarket Accessibility and Neighborhood Need in Edmonton, Canada." *Professional Geographer* 58:307–26.

Sparks, A. L., N. Bania, and L. Leete. 2010. "Comparative Approaches to Measuring Food Access in Urban Areas: The Case of Portland, Oregon." *Urban Studies*. doi:10.1177/0042098010375994.

Stephenson, G., and L. Lev. 2004. "Common Support for Local Agriculture in Two Contrasting Oregon Communities." *Renewable Agriculture and Food Systems* 19:210–17.

Stevenson, G. W., and R. Pirog. 2008. "Values-Based Supply Chains: Strategies for Agrifood Enterprises of the Middle." In *Food and the Mid-Level Farm: Renewing an Agriculture of the Middle*, edited by T. A. Lyson, G. W. Stevenson, and R. Welsh, 119–43. Cambridge, MA: MIT Press.

Tamis, L. P. 2009. *Ripe for Investment: Refocusing the Food Desert Debate on Smaller Stores, Wholesale Markets and Regional Distribution Systems*, 1–91. Cambridge: Massachusetts Institute of Technology.

Tangires, H. 2003. *Public Markets and Civic Culture in Nineteenth Century America*, 5–94. Baltimore: Johns Hopkins University Press.

———. 2008. *Public Markets*. New York: W. W. Norton.

Thilmany, D., N. McKenney, D. Mushinski, and S. Weiler. 2005. "Beggar-Thy-Neighbor Economic Development: A Note on the Effect of Geographic

Interdependencies in Rural Retail Markets." *Annals of Regional Science* 39:593–605.

Trauger, A. 2009. "Social Agency and Networked Spatial Relations in Sustainable Agriculture." *Area* 41:117–28.

Tropp, D., E. Ragland, and J. Barham. 2008. *Supply Chain Basics: The Dynamics of Change in the U.S. Food Marketing Environment.* Agriculture Handbook 728-3. US Department of Agriculture, Agricultural Marketing Service.

USDA (US Department of Agriculture). 2011. "Know Your Farmer, Know Your Food." http://www.usda.gov/wps/portal/usda/knowyourfarmer?navid=KNOWYOURFARMER.

USDA ERS (US Department of Agriculture Economic Research Service). 2012. "Chart: Farmers' Markets Concentrated in Metro Counties." *Amber Waves: The Economics of Food, Farming, Natural Resources and Rural America* 10 (4).

Ver Ploeg, M., D. Nulph, and R. Williams. 2011. "Mapping Food Deserts in the U.S." *Amber Waves: The Economics of Food, Farming, Natural Resources and Rural America* 9:46–49. http://www.ers.usda.gov/amber-waves/2011-december/data-feature-mapping-food-deserts-in-the-us.aspx#.VwPu9fkrKUk.

Watson, O. W., and A. Kwan. 2002. *The Changing Models of Inner City Grocery Retailing*, 1–52. Initiative for a Competitive Inner City. http://www.icic.org/ee_uploads/publications/TheChangingModels-02-July.pdf.

Widener, M. J., S. S. Metcalf, and Y. Bar-Yam. 2011. "Dynamic Urban Food Environments: A Temporal Analysis of Access to Healthy Foods." *AMEPRE* 41:439–41. doi:10.1016/j.amepre.2011.06.034.

Williams, T. 2013. "For Shrinking Cities, Destruction Is a Path to Renewal." *New York Times*, November 12.

Wrigley, N. 2002. "'Food Deserts' in British Cities: Policy Context and Research Priorities." *Urban Studies* 39:2029–40. doi:10.1080/0042098022000011344.

Zenk, S. N., A. J. Schulz, B. A. Israel, S. A. James, S. Bao, and M. L. Wilson. 2005. "Neighborhood Racial Composition, Neighborhood Poverty, and the Spatial Accessibility of Supermarkets in Metropolitan Detroit." *American Journal of Public Health* 95 (4): 660–67.

SECTION FIVE: COMMUNITY HEALTH AND POLICY PERSPECTIVES

Ackerman, K. 2011. *The Potential for Urban Agriculture in New York City: Growing Capacity, Food Security, and Green Infrastructure.* New York: Urban Design Lab, Earth

Institute, Columbia University.

Alaimo, K., E. Packnett, R. A. Miles, and D. J. Kruger. 2008. "Fruit and Vegetable Intake among Urban Community Gardeners." *Journal of Nutrition Education and Behavior* 40 (2): 94–101. doi:10.1016/j.jneb.2006.12.003.

Alkon, A. H., and J. Agyeman, eds. 2011. *Cultivating Food Justice: Race, Class and Sustainability*. Cambridge, MA: MIT Press.

Altman, L., L. Barry, M. Barry, K. Kühl, P. Silva, and B. Wilks. 2014. *Five Borough Farm II: Growing the Benefits of Urban Agriculture in New York City*. New York: Design Trust for Public Space.

Armstrong, D. L. 2000a. "A Community Diabetes Education and Gardening Project to Improve Diabetes Care in a Northwest American Indian Tribe." *Diabetes Educator* 26 (1): 113–20.

———. 2000b. "A Survey of Community Gardens in Upstate New York: Implications for Health Promotion and Community Development." *Health and Place* 6:319–27.

Austin, E. N., Y. A. M. Johnston, and L. L. Morgan. 2006. "Community Gardening in a Senior Center: A Therapeutic Intervention to Improve the Health of Older Adults." *Therapeutic Recreation Journal* 40 (1): 48.

Barling, D., T. Lang, and M. Caraher. 2002. "Joined-Up Food Policy? The Trials of Governance, Public Policy and the Food System." *Social Policy and Administration* 36 (6): 556–74.

BC Ministry of Health. 2011. *Evidence Review: Food Security*.

Beckie, M., and E. Bogdan. 2010. "Planting Roots: Urban Agriculture for Senior Immigrants." *Journal of Agriculture, Food Systems, and Community Development* 1 (2): 77–89.

Bellows, A., K. Brown, and J. Smit. 2005. *Health Benefits of Urban Agriculture*. Paper from the Members of the Community Food Security Coalition's North American Initiative on Urban Agriculture, February 25, 1–27. http://foodsecurity.org/pubs.html.

Blair, D., C. Giesecke, and S. Sherman. 1991. "A Dietary, Social and Economic Evaluation of the Philadelphia Urban Gardening Project." *Journal of Nutrition Education* 23:161–67.

Bradley, K., and R. E. Galt. 2013. "Practicing Food Justice at Dig Deep Farms & Produce, East Bay Area, California: Self-Determination as a Guiding Value and Intersections with Foodie Logics." *Local Environment*, May 23, 1–15. doi:10.1080/13549839.2013.790350.

Branas, C. C., R. A. Cheney, J. M. MacDonald, V. W. Tam, T. D. Jackson, and T.

R. Ten Have. 2011. "A Difference-in-Differences Analysis of Health, Safety, and Greening Vacant Urban Space." *American Journal of Epidemiology* 174 (11): 1296–306.

Brauer, M., C. Reynolds, and P. Hystad. 2013. "Traffic Related Air Pollution and Health in Canada." *Canadian Medical Association Journal* 185 (18): 1557–58.

Brinkley, C., and D. Vitiello. 2013. "From Farm to Nuisance: Animal Agriculture and the Rise of Planning Regulation." *Journal of Planning History* 13 (2): 113–35. doi:10.1177/1538513213507542.

Broad Leib, E. 2012. *Good Laws, Good Food: Putting Local Food Policy to Work for our Communities*. Jamaica Plain, MA: Harvard Law School Food Law and Policy Clinic, Community Food Security Coalition. http://blogs.law.harvard.edu/foodpolicyinitiative/files/2011/09/FINAL-LOCAL-TOOLKIT2.pdf.

Canning, P., A. Charles, S. Huang, K. R. Polenske, and A. Waters. 2010. *Energy Use in the U.S. Food System*. ERR-94. US Department of Agriculture, Economic Research Service.

Center for Active Design. 2013. "Active Design Guidelines." http://centerforactivedesign.org/guidelines/.

Center for a Livable Future. 2015. "Food Policy Council Directory." Johns Hopkins Bloomberg School of Public Health. Accessed March 25, 2016. http://www.jhsph.edu/research/centers-and-institutes/johns-hopkins-center-for-a-livable-future/projects/FPN/directory/index.html.

City of New York. 2006. Department of Housing Preservation and Development. "New Housing New York Legacy Project Request for Proposals." Accessed January 20, 2012. http://www.aiany.org/NHNY/rfp/index.php.

———. 2012. NYC Planning. "Zone Green Text Amendment Approved!" http://www.nyc.gov/html/dcp/html/greenbuildings/index.shtml.

City of Vancouver. 2014. "Rezoning Policy for Sustainable Large Developments." Accessed March 25, 2016. http://former.vancouver.ca/commsvcs/BYLAWS/bulletin/R019.pdf.

Cohen, N. 2012. "Planning for Urban Agriculture: Problem Recognition, Policy Formation, and Politics." In *Sustainable Food Planning: Evolving Theory and Practice*, edited by A. Viljoen and J. S. C. Wiskerke. Wageningen, Netherlands: Wageningen University Press.

Cohen, N., and K. Reynolds. 2014. "Resource Needs for a Socially Just and Sustainable Urban Agriculture System: Lessons from New York City." *Renewable Agriculture and Food Systems*. doi:10.1017/S1742170514000210.

Cohen, N., K. Reynolds, and R. Sanghvi. 2012. *Five Borough Farm: Seeding the Future* of *Urban Agriculture in New York City*. http://www.fiveboroughfarm.org/pdf/5BF_publication_low.pdf.

Cohen, N. and K. Wijsman. 2014. "Urban Agriculture as Green Infrastructure: The Case of New York City." *Urban Agriculture Magazine* 27:16–19. http://www.ruaf.org/publications/urban-agriculture-magazine-english-0.

Colasanti, K., C. Litjens, and M. W. Hamm. 2010. *Growing Food in the City: The Production Potential of Detroit's Vacant Land*. Report by the C. S. Mott Group for Sustainable Food Systems, Michigan State University, July 13, 1–13.

Connelly, S., S. Markey, and M. Roseland. 2011. "Bridging Sustainability and the Social Economy: Achieving Community Transformation through Local Food Initiatives." *Critical Social Policy* 31 (2): 308–24.

Cronbach, L. J., and P. E. Meehl. 1955. "Construct Validity in Psychological Tests." *Psychological Bulletin* 52 (4): 281.

Dahlberg, K., K. Clancy, R. L. Wilson, and J. O'Donnell. 2002. "Local Food Policy Goals and Issues." In *Strategies, Policy Approaches, and Resources for Local Food System Planning and Organizing: A Resource Guide*. Minnesota Food Association. http://unix.cc.wmich.edu/~dahlberg/F1.pdf.

Day Farnsworth, L. Forthcoming. "Beyond Policy: Race, Class, Leadership and Agenda-Setting in North American Food Policy Councils." In *Leadership, Learning, and Food: Global Perspectives on Food Systems Transformation*, edited by C. Etmanski. Dordrecht, Netherlands: Sense Publishers.

Dodman, D. 2009. "Blaming Cities for Climate Change? An Analysis of Urban Greenhouse Gas Emissions Inventories." *Environment and Urbanization* 21 (1): 185–201.

Elder, R. F. 2005. "Protecting New York City's Community Gardens." *New York University Environmental Law Journal* 13:769. http://heinonlinebackup.com/hol-cgi-bin/get_pdf.cgi?handle=hein.journals/nyuev13&section=21.

Feenstra, G., S. McGrew, and D. Campbell. 1999. *Entrepreneurial Community Gardens: Growing Food, Skills, Jobs, and Communities*. Report no. 21587. University of California Agricultural and Natural Resources Publication.

Fisher, A. 1999. *Hot Peppers and Parking Lot Peaches: Evaluating Farmers' Markets in Low Income Communities*. Report by the Community Food Security Coalition, December 21, 1–66.

Flynn, K. 1999. *An Overview of Public Health and Urban Agriculture: Water, Soil, and Crop Contamination and Emerging Urban Zoonoses*. Cities Feeding People Report 30.

Food and Agriculture Organization. 2012. *State of Food Insecurity in the World 2012*. United Nations.

Francis, C., G. Lieblein, H. Steinsholt, T. A. Breland, J. Helenius, N. Sriskandarajah, and L. Salomonsson. 2005. "Food Systems and Environment: Building Positive Rural-Urban Linkages." *Human Ecology Review* 12 (1): 60–71. http://www.humanecologyreview.org/pastissues/her121/francisetal.pdf.

Garvin, E. C., C. C. Cannuscio, and C. C. Branas. 2013. "Greening Vacant Lots to Reduce Violent Crime: A Randomised Controlled Trial." *Injury Prevention* 19 (3): 198–203.

Gatto, N. M., E. E. Ventura, L. T. Cook, L. E. Gyllenhammer, and J. N. Davis. 2012. "LA Sprouts: A Garden-Based Nutrition Intervention Pilot Program Influences Motivation and Preferences for Fruits and Vegetables in Latino Youth." *Journal of the Academy of Nutrition and Dietetics* 112 (6): 913–20.

Glover, T., K. Shinew, and D. Parry. 2005. "Association, Sociability, and Civic Culture: The Democratic Effect of Community Gardening." *Leisure Sciences* 27 (1): 75–92. doi:10.1080/01490400590886060.

Golden, S. 2013a. *Annotated Bibliography: Economic, Social, and Health Impacts of Urban Agriculture*. Compiled for UC Division of Agricultural and Natural Resources. http://ucanr.edu/sites/UrbanAg/files/185843.pdf.

———. 2013b. *Urban Agriculture Impacts at a Glance*. Compiled for UC Division of Agricultural and Natural Resources. http://asi.ucdavis.edu/programs/sarep/publications/food-and-society/uaspreadsheet-2013.pdf.

——— 2013c. *Urban Agriculture Impacts: Social, Health, and Economic, a Literature Review*. Compiled for UC Division of Agricultural and Natural Resources. http://asi.ucdavis.edu/programs/sarep/publications/food-and-society/ualitreview-2013.pdf.

Goldstein, M., J. Bellis, S. Morse, A. Myers, and E. Ura. 2011. *Urban Agriculture: A Sixteen City Survey of Urban Agriculture Practices across the Country*. Survey written and compiled by Turner Environmental Law Clinic at Emory University Law School and Georgia Organics, November 1, 1–94.

Green, M., H. Moore, and J. O'Brien. 2006. *When People Care Enough to Act: ABCD in Action*. Toronto, ON: Inclusion Press.

Harper, A., A. Shattuck, E. Holt-Giménez, A. Alkon, and F. Lambrick. 2009. *Food Policy Councils: Lessons Learned*. Oakland, CA: Food First Institute for Food and Development Policy.

Healey, P. 2012. "Re-enchanting Democracy as a Mode of Governance." *Critical Policy Studies* 6 (1): 19–39.

Hendrickson, M. K., and M. Porth. 2012. *Urban Agriculture: Best Practices and Possibilities*. Columbia: University of Missouri Extension, July 2, 1–52.

Herman, D. R., G. G. Harrison, A. A. Afifi, and E. Jenks. 2008. "Effect of a Targeted Subsidy on Intake of Fruits and Vegetables among Low-Income Women in the Special Supplemental Nutrition Program for Women, Infants, and Children." *American Journal of Public Health* 98 (1): 98–105. doi:10.2105/AJPH.2005.079418.

Hodgson, K. 2012. *Planning for Food Access and Community-Based Food Systems: A National Scan and Evaluation of Local Comprehensive and Sustainability Plans.* American Planning Association Report, November 26, 1–175.

Hodgson, K., M. C. Campbell, and M. Bailkey. 2011. *Urban Agriculture: Growing Healthy, Sustainable Places.* PAS 563. APA American Planning Advisory Service.

Jansen, B. N., and E. S. Mills. 2013. "Distortions Resulting from Residential Land Use Controls in Metropolitan Areas." *Journal of Real Estate Finance Economics* 46:193–202.

Joshi, A., and A. M. Azuma. 2012. *Bearing Fruit: Farm to School Program Evaluation Resources and Recommendations.* National Farm to School Network and Center for Food and Justice Urban and Environmental Policy Institute, Occidental College, October 31, 1–211.

Kahneman, D., and A. B. Krueger. 2006. "Developments in the Measurement of Subjective Well-Being." *Journal of Economic Perspectives* 20 (1): 3–24.

Kerton, S., and A. J. Sinclair. 2009. "Buying Local Organic Food: A Pathway to Transformative Learning." *Agriculture and Human Values* 27 (4): 401–13. doi:10.1007/s10460-009-9233-6.

Kingsley, J. Y., M. Townsend, and C. Henderson-Wilson. 2009. "Cultivating Health and Wellbeing: Members' Perceptions of the Health Benefits of a Port Melbourne Community Garden." *Leisure Studies* 28 (2): 207–19.

Koc, M., R. MacRae, E. Desjardins, and W. Roberts. 2008. "Getting Civil about Food: The Interactions between Civil Society and the State to Advance Sustainable Food Systems in Canada." *Journal of Hunger and Environmental Nutrition* 3 (2–3): 122–44.

Krasny, M. E., and R. Doyle. 2002. "Participatory Approaches to Program Development and Engaging Youth in Research: The Case of an Intergenerational Urban Community Gardening Program." *Journal of Extension* 40 (5): 1–16.

Lackey, J. F. 1998. *Evaluation of Community Gardens.* Jill Florence Lackey and Associates, February 14, 1–98.

Landis, B., T. E. Smith, M. Lairson, K. Mckay, H. Nelson, and J. O'Briant. 2010. "Community-Supported Agriculture in the Research Triangle Region of North Carolina: Demographics and Effects of Membership on Household Food

Supply and Diet." *Journal of Hunger and Environmental Nutrition* 5 (1): 70–84. doi:10.1080/19320240903574403.

Levkoe, C. Z. 2006. "Learning Democracy through Food Justice Movements." *Agriculture and Human Values* 23:89–98. doi:10.1007/s10460-005-5871-5.

Litt, J. S., M-J. Soobader, M. S. Turbin, J. W. Hale, M. Buchenau, and J. A. Marshall. 2011. "The Influence of Social Involvement, Neighborhood Aesthetics, and Community Garden Participation on Fruit and Vegetable Consumption." *American Journal of Public Health* 101 (8): 1466–73.

Loorbach, D. 2007. *Transition Management: New Mode of Governance for Sustainable Development*. Utrecht, Netherlands: International Books.

Loorbach, D., N. Frantzeskaki, and W. Thissen. 2011. "A Transition Research Perspective on Governance for Sustainability." In *European Research on Sustainable Development*. Vol. 1, *Transformative Science Approaches for Sustainability*, edited by C. C. Jaeger, J. D. Tabara, and J. Jaeger, 73–90. Berlin: Springer.

Magnus, K., M. Matroos, and J. Strackee. 1979. "Walking, Cycling, or Gardening, with or without Interruption, in Relation to Acute Coronary." *American Journal of Epidemiology* 110 (6): 724–33.

Manuel, P., and K. Thompson. n.d. *The Role of Third Place in Community Health and Well-Being*. Centre for Urban Health Initiatives. http://www.cuhi.utoronto.ca/research/neighbourhoodrig.html.

McClintock, N. 2014. "Radical, Reformist, and Garden-Variety Neoliberal: Coming to Terms with Urban Agriculture's Contradictions." *Local Environment*, January 10, 1–25. doi:10.1080/13549839.2012.752797.

McCormack, L. A., M. Story, N. I. Larson, and M. N. Laska. 2010. "Review of the Nutritional Implications of Farmers' Markets and Community Gardens: A Call for Evaluation and Research Efforts." *Journal of the American Dietetic Association* 110 (3): 399–408. doi:10.1016/j.jada.2009.11.023.

Mendes, W., K. Balmer, T. Kaethler, and A. Rhoads. 2008. "Using Land Inventories to Plan for Urban Agriculture: Experiences from Portland and Vancouver." *Journal of the American Planning Association* 74 (4): 435–49.

Morales, A., and G. Kettles. 2009. "Zoning for Public Markets and Street Vendors." *Zoning Practice* 2, American Planning Association.

Mukherji, N., and A. Morales. 2010. "Zoning for Urban Agriculture." *Zoning Practice* 3. American Planning Association. http://www.planning.org/zoningpractice/2010/pdf/mar.pdf.

Muller, M., A. Tagtow, S. L. Roberts, and E. MacDougall. 2009. "Aligning Food Systems Policies to Advance Public Health." *Journal of Hunger and Environmental*

Nutrition 4 (3–4): 225–40.
Ober Allen, J., K. Alaimo, D. Elam, and E. Perry. 2008. "Growing Vegetables and Values: Benefits of Neighborhood-Based Community Gardens for Youth Development and Nutrition." *Journal of Hunger and Environmental Nutrition* 3 (4): 418–39. doi:10.1080/19320240802529169.
Packaged Facts. 2007. "Local and Fresh Foods in the U.S." http://www.packagedfacts.com/Local-Fresh-Foods-1421831/.
Park, S-A., C. A. Shoemaker, and M. D. Haub. 2009. "Physical and Psychological Health Conditions of Older Adults Classified as Gardeners or Nongardeners." *HortScience* 44 (1): 206–10.
Park, Y., J. Quinn, K. Florez, J. Jacobson, K. Neckerman, and A. Rundle. 2011. "Hispanic Immigrant Womens' Perspective on Healthy Foods and the New York City Retail Food Environment: A Mixed-Method Study." *Social Science and Medicine* 73 (1): 13–21. doi:10.1016/j.socscimed.2011.04.012.
Patel, I. C. 1991. "Gardening's Socioeconomic Impacts." *Journal of Extension* 29 (4): 1–3.
Philips, A. 2013. *Designing Urban Agriculture: A Complete Guide to the Planning, Design, Construction, Maintenance and Management of Edible Landscapes*. Hoboken, NJ: John Wiley and Sons.
Pothukuchi, K., and J. Kaufman. 1999. "Placing the Food System on the Urban Agenda: The Role of Municipal Institutions in Food Systems Planning." *Agriculture and Human Values* 16 (2): 213.
———. 2000. "The Food System: A Stranger to the Planning Field." *Journal of the American Planning Association* 66 (2): 113–24.
Quinn, C. 2010. *FoodWorks: A Vision to Improve NYC's Food System*. http://council.nyc.gov/downloads/pdf/foodworks_fullreport_11_22_10.pdf.
Rashleigh, J. 2013. Farmers on 57th Recommendations for the Pearson Dogwood Redevelopment. Report in possession of the author.
Reynolds, K., and N. Cohen. 2016. *Beyond the Kale: Urban Agriculture and Social Justice Activism in New York City*. Athens: University of Georgia Press.
Robert Wood Johnson Foundation. n.d. Leadership for Healthy Communities: Advancing Policies to Support Healthy Eating and Active Living. Accessed October 1, 2013. http://www.leadershipforhealthycommunities.org/index.php/cdc-makes-funding-for-obesity-prevention-available-to-states-and-communities-hiddenmenu-157.
Robinson-O'Brien, R., M. Story, and S. Heim. 2009. "Impact of Garden-Based Youth Nutrition Intervention Programs: A Review." *Journal of the American Dietetic Association* 109 (2): 273–80.

Saldivar-Tanaka, L., and M. E. Krasny. 2004. "Culturing Community Development, Neighborhood Open Space, and Civic Agriculture: The Case of Latino Community Gardens in New York City." *Agriculture and Human Values* 21:399–412.

Sanghvi, R. 2012. *Outcomes/Benefits of Urban Agriculture Supported by the Evidence Base*. Five Borough Farms. http://www.fiveboroughfarm.org/resources/.

Schukoske, J. E. 2000. "Community Development through Gardening: State and Local Policies Transforming Public Space." *Legislation and Public Policy* 3:351–92.

Sharp, J. S., E. Imerman, and G. Peters. 2002. "Community Supported Agriculture (CSA): Building Community among Farmers and Non-Farmers." *Journal of Extension* 40 (3): 1–6.

Shove, E. 2003. *Comfort, Cleanliness and Convenience*. Oxford: Berg.

Shove, E., M. Pantzar, and M. Watson. 2012. *The Dynamics of Social Practice: Everyday Life and How It Changes*. London: SAGE.

Shove, E., and G. Walker. 2010. "Governing Transitions in the Sustainability of Everyday Life." *Research Policy* 39 (4): 471–76.

Steel, C. 2008. *Hungry City: How Food Shapes Our Lives*. London: Random House.

The Stop. 2014. "Annual Report 2014." http://thestop.org/wp-content/uploads/thestop_annualreport-2014-digital.pdf.

Strom, S. 2007. "$500 Million Pledged to Fight Childhood Obesity." *New York Times*, April 4. http://www.nytimes.com/2007/04/04/health/04obesity.html?_r=0.

Suarez-Balcazar, Y. 2006. "African Americans' Views on Access to Healthy Foods: What a Farmers' Market Provides." *Journal of Extension* 44 (2): 1–7.

Sumner, J., H. Mair, and E. Nelson. 2010. "Putting the Culture Back into Agriculture: Civic Engagement, Community and the Celebration of Local Food." *International Journal of Agricultural Sustainability* 8 (1–2): 54–61.

Teig, E., J. Amulya, L. Bardwell, M. Buchenau, J. A. Marshall, and J. S. Litt. 2009. "Collective Efficacy in Denver, Colorado: Strengthening Neighborhoods and Health through Community Gardens." *Health and Place* 15 (4): 1115–22.

Thibert, J. 2012. "Making Local Planning Work for Urban Agriculture in the North American Context: A View from the Ground." *Journal of Planning Education and Research* 32 (3): 349–57.

Thompson, C. J., K. Boddy, K. Stein, R. Whear, J. Barton, and M. H. Depledge. 2011. "Does Participating in Physical Activity in Outdoor Natural Environments Have a Greater Effect on Physical and Mental Wellbeing Than Physical Activity Indoors? A Systematic Review." *Environmental Science and Technology* 45 (5): 1761–72.

Travaline, K., and C. Hunold. 2010. "Urban Agriculture and Ecological Citizenship in Philadelphia." *Local Environment* 15 (6): 581–90. doi:10.1080/13549839.2010.487529.

Tucs, E., and B. Dempster. 2007. *Linking Health and the Built Environment: A Literature Review*. Ontario Healthy Communities Coalition. http://www.ohcc-ccso.ca/en/linking-health-and-the-built-environment-a-literature-review.

Twiss, J., J. Dickinson, S. Duman, T. Kleinman, H. Paulsen, and L. Rilveria. 2003. "Community Gardens: Lessons Learned from California Healthy Cities and Communities." *American Journal of Public Health* 93 (9): 1435–38.

USDA (US Department of Agriculture). 2010. "Getting to Scale with Regional Food Hubs." http://blogs.usda.gov/2010/12/14/getting-to-scale-with-regional-food-hubs/.

US EPA (US Environmental Protection Agency). 2014. *Municipal Solid Waste Generation, Recycling, and Disposal in the United States: Facts and Figures for 2012*. https://www3.epa.gov/epawaste/nonhaz/municipal/msw99.htm.

US HUD (Department of Housing and Urban Development). 2011. *Final PHA Plan: Annual Plan for Fiscal Year 2012*. http://www.nyc.gov/html/nycha/downloads/pdf/FY2012-AnnualPlan.pdf.

Wakefield, S., F. Yeudall, C. Taron, J. Reynolds, and A. Skinner. 2007. "Growing Urban Health: Community Gardening in South-East Toronto." *Health Promotion International* 22 (2): 92–101.

Wannamethee, S. G., and A. G. Shaper. 2001. "Physical Activity in the Prevention of Cardiovascular Disease." *Sports Medicine* 31 (2): 101–14.

Warburton, D. E. R., C. W. Nicol, and S. S. D. Bredin. 2006. "Health Benefits of Physical Activity: The Evidence." *Canadian Medical Association Journal* 174 (6): 801–9.

Warde, A. 2005. "Consumption and Theories of Practice." *Journal of Consumer Culture* 5 (2): 131–53.

Wekerle, G. R. 2004. "Food Justice Movements: Policy, Planning, and Networks." *Journal of Planning Education and Research* 23 (4): 378–86.

Weltin, A. M., and R. P. Lavin. 2012. "The Effect of a Community Garden on HgA1c in Diabetics of Marshallese Descent." *Journal of Community Health Nursing* 29 (1): 12–24.

White, M. M. 2010. "Shouldering Responsibility for the Delivery of Human Rights: A Case Study of the D-Town Farmers of Detroit." *Race/Ethnicity* 3 (2): 189–211. doi:10.2979/RAC.2010.3.2.189.

World Health Organization. 2014. *Diet and Physical Activity: A Public Health Authority*. http://www.who.int/dietphysicalactivity/en.

Zick, C. D., K. R. Smith, L. Kowaleski-Jones, C. Uno, and B. J. Merrill. 2013. "Harvesting More Than Vegetables: The Potential Weight Control Benefits of Community Gardening." *American Journal of Public Health* 103 (6): 1110–15.

INDEX

African Americans, 29, 196, 198, 202
agriculture: colonial, 42; rooftop, 117–118, 217, 221, 224–227 passim, 279; suburban, 41–57 passim
agrihood, 277–278
agroecology, 107–124 passim
agronomic system, 109
American Indians. *See* Native Americans
aquaponics, 112

Beacon Food Forest, 277
built environment, 265–288 passim

Capital Regional District (Vancouver Island), 274
Cascadia, 59–81 passim
Chamberlin, Emily, 35
Chicago Transit Authority (CTA), 199
Community Commons, 240
Community Food Security Coalition, 232
Community Services Unlimited, 121
community supported agricultural (CSA) program, 67, 233
community-based organization (CBO), 198, 286
composting, 67–68, 83–103 passim, 110, 288; regulation, 49–50, 55, 110; vermicomposting, 110
Comprehensive Farm Review, 102
cooking, 14, 122, 130–139 passim, 204, 216, 243, 276
cooperative extension, 113, 123
cooperatives, farm, 109–110
cooperatives, food (retail), 18, 78. *See also* Producer's Cooperative Market Association
Craig, Charles, 14–15
crop production, 72, 73, 111–124 passim; crop nutrients, 96, 110; organic production, 281; season extension, 115; soil quality, 109–111, 117. *See also* gardens; gardening
cultural capital, 129
culture, 136, 213. *See also* food: culturally appropriate; food: traditional

Department of Housing Preservation and Development (HPD), 224–225, 227
diet, 128–129, 139, 194–195, 225, 231–234. *See also* cooking
Duluth Commercial Club, 13–16
Duluth Community Garden Program, 18
Duluth Farmers Market, 17
Duluth Homecroft Association (DHA), 14

Eat Oregon First, 276
economics, 119–120
edible landscaping, 67, 226, 270–271, 275, 277–278
education: classes/courses, 114–115, 152, 204, 217, 243; educational trains, 15
Electronic Benefit Transfer (EBT) machines, 252
Elijah's Promise, 150, 153–155
Elson, William H., 28
environment: emissions, 63, 83, 86, 268, 281; microclimates, 115, 123; pollution, 84, 90, 221, 225–226, 281. *See also* New York City: Department of Environmental Protection (DEP)
equitability, 108, 120–122
Eva, Major, 14
extension. *See* cooperative extension

Falling Fruit, 277
farmers markets: EBT/SNAP use, 202–204, 252; food access, 128, 193–204, 272, 274; fostering community, 236; healthy food consumption, 233, 291; history, 21–37 passim; increase in numbers, 249; street design, 277; vendor access, 185–188. *See also* Duluth Farmers Market; Grand Rapids Farmers Market; New Brunswick Community Farmers Market; Rutgers Gardens Farmers Market
farming: corporate, 17; vertical, 117–118. *See also* farms
farms: peri-urban, 177, 291
Federal Way (Washington State), 49, 52–55
feminists, first-wave, 27
Five Borough Farm Project, 232–233, 239–240
food: culturally appropriate, 122, 135–139, 206, 236; insecurity, 8, 144, 178, 192–210 passim, 274; safety, 109, 114, 123, 253, 288; traditional, 127, 129; unhealthy, 233–234
food desert, 128, 192–210, 240
Food Distribution Program, 129
food pantries, 121, 131–133, 139
food stamps. *See* Supplemental Nutrition Assistance Program (SNAP)
food street, 277
food systems framework, 6–7
FoodWorks, 219, 221
Forterra, 53
funding: block grant, 143–150; grant, 75–76, 120, 231, 246, 249; public, 29, 35, 55, 81, 249

gardening: container, 46, 105, 111–112, 127, 130–131, 138–139; guerilla, 218; home, 28–29; public, 22–23, 27–32; rooftop, 46, 53, 279; vacant lot, 4, 21, 48, 109, 162, 217, 237, 241, 279, 290; windowsill, 131. *See also* gardens; Gardens for Healthy Communities

gardens: for-profit, 47, 55; market, 48–49, 67, 282; raised-bed, 111, 122, 150–151, 269; school, 27–30, 65, 80, 242, 279; urban, 126–127, 139, 141, 219; Victory Garden, 30. *See also* gardening; Gardens for Healthy Communities; New York City Community Garden Coalition
Gardens for Healthy Communities, 219
Grand Rapids Federation of Women's Clubs (GRFWC), 32–33
Granville Island (Vancouver, British Columbia), 280
Green Barn Market, 276
green infrastructure, 60, 224–226
greenhouses, 63, 113–117 passim, 152, 220–221, 224, 256
Greysolon Farms Company, 15
grocery stores. *See* supermarkets
Growing Power, 11, 121, 262, 291

Hamilton, Eva McCall, 21–22, 28, 29, 33, 35
health: determinants, 238; mental health, 44; obesity, 45, 128, 144, 196, 234, 241–242, 249, 265, 272; outcomes, 127–128, 139, 194, 204, 231–238 passim, 268, 271; pathways, 237–244 passim; physical activity, 234, 241, 249, 266–267; whole health, 280–281. *See also* Health Impact Assessment (HIA); National Collaborative on Childhood Obesity Research; public health
Health Impact Assessment (HIA), 239–240
Homeboy Industries, 122
homeowners association, 45–46, 52
hoop houses, 152–153
Hostetter, A. B., 14–15

immigrants, 24, 29, 43, 161, 194, 198, 223, 236
Indian Center, 130–138 passim
insect management. *See* pest management

Johnson & Johnson Corporation, 143–156 passim

land lease, 47, 72–73, 74–75
Land Stewardship Project, 18
Leonard, Charles H., 25–26
Lindeman, E. C., 30
Link Card. *See* Supplemental Nutrition Assistance Program (SNAP)
livestock, 77, 111–112, 114, 184, 186
local food movement, 96, 278–279

master gardener program, 113
Miller, Louise Klein, 30
mobile market, 250, 270–271, 274

National Collaborative on Childhood Obesity Research, 241–242
National Farm to School Network, 242
Native Americans, 105, 127–139 passim. *See also* Indian Center
New Brunswick Community Farmers Market, 141–156 passim
New York City: City Community Garden Coalition, 219–220; Department of Environmental Protection (DEP), 225–226; Housing Authority (NYCHA), 224–225
New York Restoration Project (NYRP), 219
North American Industry Classification

Index   331

System (NAICS), 193, 208
Northeast Demonstration Farm and
    Experimental Station, 16
Nutrition Environment Measures Study
    (NEMS), 208–209

obesity. *See under* health
Operation GreenThumb, 218–219
orchard, community, 50, 63, 80, 275,
    277, 279

Pearson Dogwood, 280–283
pest management, 114–116, 120
pollution. *See under* environment
poverty, 45, 122–123, 128, 144
Producer's Cooperative Market
    Association, 16–17
public health, 291–292
Public Health–Seattle and King County,
    51–52
public retail market, 21–27 passim,
    32–37 passim. *See also* farmers
    market
Putting Prevention to Work (CPPW),
    51, 53

recycling, 84–86, 90–96 passim
Red Tomato, 276
redlining. *See* supermarket redlining
Regional Access, LLC (RA), 177–178,
    183–189, 291
regulation: Cascadia example, 59–82;
    centrality to food production,
    8; composting regulations,
    89–92, 96–103; decentralization,
    246; future policy takeaways,
    55–57; history, 42–44, 288; impact,
    113–114; influence, 286–287; King
    County, WA, example 51–55; policy
    attitudes, 76–79; research need,

289–290; Royal Oak, MI, example,
    48–50; urban farms, 118–119; Wheat
    Ridge, CO, example, 50–51; zero-
    waste resolutions, 93–94. *See also*
    zoning
Resource Conservation and Recovery
    Act (RCRA), 89–90
retail market. *See* public retail market
Riverpark Farm, 222
Robertson, Gregor, 64
Royal Oak (Michigan), 48–50, 54
Rutgers Gardens Farmers Market, 144
Rutgers University, 143–156 passim

sanitation. *See under* waste
Shaw's Tripartite Approach, 206
Snoqualmie (Washington State), 49,
    52, 54
stacked functions, 68
The Stop: Community Food Centre, 276
supermarket redlining, 193, 196
supermarkets, 22, 33, 133, 161, 178,
    180, 191–196, 207, 270
Supplemental Nutrition Assistance
    Program (SNAP), 148–149,
    202–204, 252
supply chains, 178–183 passim, 193
Sustainable Farming Association, 18

Thompson, Mark J., 16
Tilma, George P., 32–33
traditional foods. *See under* food
transportation, 115, 128, 177, 195,
    206–207, 266, 270, 272, 281. *See also*
    Chicago Transit Authority (CTA)
Trust for Public Land (TPL), 219

University of California's Division of
    Agriculture and Natural Resources,
    232

University of Minnesota, 16
University of Wisconsin–Madison Community and Regional Food Systems Project, 247
urban blight, 117–118, 121, 237, 290
urbanization, 127
US Environmental Protection Agency (EPA), 83, 99

Van Buren, Frances, 29–30
vermicomposting. *See under* composting
Victory Garden. *See under* gardens
volunteer labor, 73–74, 81, 113, 116, 137
vouchers. *See* Supplemental Nutrition Assistance Program (SNAP)

waste: food, 85–86, 99, 103, 110, 216, 272; management, 42, 50, 83–103 passim; sanitation, 43; source separated, 92–93, 100; zero waste, 89, 93–95
water: access, 76, 114, 118; reclaimed, 288–289; stormwater, 218, 221–222, 226–227
Wheat Ridge (Colorado), 50–51, 54
wholesale: distribution, 33, 177, 182, 188, 270, 278; markets, 23–28 passim, 118, 180
Williams, John G., 15
Women, Infants, and Children (WIC), 148–149, 233
women's clubs, 21–37 passim
World War II, 17

YMCA, 14

zero waste. *See under* waste
zoning: composting, 91–92, 96–101; food access, 139; history, 267; King County, WA, example, 53–54; Portland, OR, example, 63, 78–79, 273–274; revisions, 230; rezoning in Vancouver, 275; Royal Oak, MI, example, 48–50; urban farms, 118–119; Wheat Ridge, CO, example, 50–51. *See also* regulations